Principles of Quality Management

Principles of
Quality Management

Edited by **Stacy Escobar**

LANRYE
INTERNATIONAL

New Jersey

Published by Clanrye International,
55 Van Reypen Street,
Jersey City, NJ 07306, USA
www.clanryeinternational.com

Principles of Quality Management
Edited by Stacy Escobar

International Standard Book Number: 978-1-63240-418-3 (Hardback)

Printed in the United States of America.

Contents

Preface

An in-depth discussion regarding quality management and its practices has been highlighted in this up-to-date book. It consists of a compilation of reviews and research works contributed by professionals from across the globe. A practical approach to quality management will facilitate the readers with comprehensive information regarding topics ranging from basic to total quality practices in organizations, providing a systematic coverage of topics. The primary focus of this book is on quality management practices in organizations and dealing with particular total quality practices to quality management systems. This book can be used as a valuable source of reference at colleges, universities, corporate organizations, and for individual readers who wish to increase their knowledge regarding this field. The information provided in this book will serve as a helpful and useful guide for practitioners seeking to comprehend and use suitable techniques for implementation of total quality.

This book unites the global concepts and researches in an organized manner for a comprehensive understanding of the subject. It is a ripe text for all researchers, students, scientists or anyone else who is interested in acquiring a better knowledge of this dynamic field.

I extend my sincere thanks to the contributors for such eloquent research chapters. Finally, I thank my family for being a source of support and help.

Editor

Section 1

Quality Concepts and Practices

Quality Management System and Practices

Ng Kim-Soon
*Universiti Tun Hussein
Onn Malaysia
Malaysia*

1. Introduction

Quality is a perceptual, conditional and somewhat subjective attribute of a product or service. Its meaning in business has developed over time. It has been understood differently and interpreted differently by different people. A business will benefit most through focusing on the key processes that provide their customers with products and services. Producers may measure the conformance quality, or degree to which the product or service was made according to the required specification. Customers on the other hand, may focus on the quality specification of a product or service, or compared it with those that are available in the marketplace. In a modern global marketplace, quality is a key competency which companies derive competitive advantage. Achieving quality is fundamental to competition in business in propelling business into new heights.

Many quality management philosophies, methodologies, concepts and practices were created by quality gurus to manage quality of product and service in an organization. These practices have evolved over time to create sustainable sources of competitive advantage. New challenges faced by managers are addressed to improve organization's performance and future competition. In the total quality management form, it is a structured management system adopted at every management levels that focused on ongoing effort to provide product or service. Its integration with the business plan of the organization can exact positive influence on customer satisfaction and organizational performance.

This chapter dealt with what is quality and TQM, cost of quality, linking quality management system to organizational performance, its impact on organizations and approaches of implementing TQM and the quality journey.

2. What is quality and TQM?

The business meanings of quality have developed over time. Among the interpretations of quality stated by Wikipedia from the various sources is tabled as below:

	Author / Authority/Source	Definition
1.	*ISO 9000:2008*, International Organization for Standardization	Degree to which a set of inherent characteristics fulfills requirements
2.	Six Sigma (Motorola University)	Number of defects per million opportunities
3.	Subir Chowdhury (2005)	Quality combines people power and process power
4.	Philip B. Crosby (1979)	Conformance to requirements
5.	Joseph M. Juran (American Society for Quality)	Fitness for use
6.	Genichi Taguchi (1992)	Uniformity around a target value and the loss a product imposes on society after it is shipped
7.	Peter Drucker (1985)	Quality in a product or service is not what the supplier puts in. It is what the customer gets out and is willing to pay for
8.	Edwards Deming (1986) and Walton, Mary and Edwards Deming (1988)	The efficient production of the quality that the market expects, and costs go down and productivity goes up as improvement of quality is accomplished by better management of design, engineering, testing and by improvement of processes

Table 1. Interpretations of Quality

The development of TQM can be traced to several consultants including Crosby, Juran, Taguchi and Deming. They showed different ways in defining quality. Crosby stresses on zero defects programme through process improvement to pursuing conformance to customers requirements. Juran and Deming stress primarily on leadership qualities, management commitment and involvement to achieve quality goals. Juran emphasizes symptom cause while and Deming emphasizes on 14 quality points. Taguchi emphasizes on concept that any deviation from the required specification results in loss and that organization need to strive to determine and meet customer's specification. In ISO 9000, it emphasizes on the need of good documentations, traceability and records keeping.

In managing quality, the focus is not only on quality of product and service itself. It is also on the means to achieve it. Thus, quality management uses management techniques and tools in quality assurance and control of processes to achieve consistent quality of products and services. Many definitions of quality can be found in the TQM literature (Kaynak, 2003, Shah and Ward 2003, Prajogo and Brown 2004, Prajogo and Sohal 2006, and Ahire, 1997). Feigenbaum (1991) defines TQM as an effective system for integrating the quality development, quality maintenance and quality improvement efforts of the various groups in an organization so as to enable production and service at the most economical levels, which allows for full customer satisfaction. It is an integrative philosophy of management practice in an organization for continuous improvement of their product, services and processes. It capitalizes on the involvement of management, employees, suppliers, and its customers to meet or exceed customer satisfactions and expectations. Review by Cua, McKone, and Schroeder (2001) found that there are nine areas of common TQM practices in organizations. These are cross-functional product design, process management, supplier quality management, customer involvement, information and feedback, committed leadership, strategic planning, cross-functional training, and employee involvement. Saraph, Benson, and Schroeder (1989) identified eight areas of primary quality concerns which they called critical quality factors. These factors of quality management were compiled from an exhaustive review of articles, books and studies of eminent quality practitioners and academics. These are the role of management leadership and quality policy; training; process management; employee relations; product / service design; supplier quality management; the role of the quality department; and quality data and reporting. They also identified that managerial commitment to quality combines several functions as one of the vital imperatives for the success of any quality improvement programme.

3. Cost of quality

Good quality product or service enables an organization to attract and retain customers. Poor quality leads to dissatisfaction to customers. As such, the costs of poor quality are not just those of immediate waste or rework and rectification, it is also the loss of future sales and subsequently the organization performance. The concept of quality costs was first described by Feigenbaum (1956) as a mean to quantify the total cost of quality-related efforts and deficiencies. According to Feigenbaum, total quality costs come from prevention costs, appraisal costs and quality failure costs. The prevention and appraisal costs are the cost of conformance. Examples of prevention cost are activities in quality planning, statistical process control, investment in quality-related information systems, quality training and workforce development, product-design verification, and systems development and management. Examples that involve appraisal costs include activities in testing and inspection of purchased materials, acceptance testing, inspection, testing, checking labor, setup for test or inspection, test and inspection equipment, quality audits and field testing.

Prevention costs arise from efforts to keep defects from occurring at all, while appraisal costs arise from detecting defects via inspection, test, and audit. On the other hand, cost of non-conformance is the quality failure costs which are comprised of internal quality failure and external quality failure. Internal failure costs arise from defects caught internally and dealt with by discarding or repairing the defective items, while external failure costs arise from defects that actually reach customers. Prevention of poor quality will reduce quality failure costs. It can also lead to the reduction of cost due to the need for many non-value added inspection and appraisal activities costs. Thus, organization needs to evaluate the costs involved of operating an effective quality management system to ensure an effective cost effective. In other words, it is crucial for organization to determine the trade-off between prevention, appraisal and failure costs for a specified level of quality performance for cost effectiveness so that investments in quality is based on cost improvement and profit enhancement. Thus, quality costs can serve as a means to measure, analyze, budget, and predict (Feigenbaum, 1991).

According to Smith (1998), the emphasis on cost, quality and time has generated many management changes with significant accounting implications with implementation of strategic initiative such as the use of activity based costing (ABC) on TQM initiative. ABC is perfectly suited to TQM because it encourages management to analyze activities and determine their value to the customer (Steimer, 1990). Shepard (1995) suggested that an economics of quality approach can be integrated with ABC for strategic cost effectiveness. Many companies found that ABC aligned well with TQM processes (Anderson and

Sedatole, 1998). Evidence of the context-specific benefit of TQM and ABC was found in case studies performed by Cooper, Kaplan, Maisel and Oehm (1992) where all five manufacturing companies studied found ABC and TQM to be highly compatible and mutually supporting. ABC and other strategic business initiatives complement and enhance each other, rather than being individually necessary and sufficient conditions for improvement (Anderson, 1995).

4. Linking quality management system to organizational performance

Sila (2005) reviewed that research works have often link TQM practices with multidimensional measures organizational performance of both financial and non-financial measures. Kaynak (2003) reported that TQM practices can directly affect financial performance, it also affect indirectly on increasing market competitiveness (Chong and Rundus, 2004), innovation (Singh and Smith, 2004), and productivity (Rahman and Bullock, 2005).

Empirical evidence supports the argument that by focusing on quality, an organization can substantially improve its performance (Peters and Austin, 1985 and ; Yahya, Salleh and Keat, 2001). Literature reviews by Lewis, Pun and Lalla (2006) on studies conducted in different countries including Costa Rica, Thailand, Indonesia, Palestine, Singapore, Australia, China and Hong Kong indicated that there are 12 criteria for successful implementation of quality management system. These are quality data and reporting; customer satisfaction, human resource utilization; management and process quality; management commitment; continuous improvement; leadership; strategic quality planning; performance measurement; customers focus; and contact with suppliers and professional associates. Literature investigation of TQM studies published between 1989 and 2000 by Sila and Ebrahimpour (2002) found 25 critical quality factors most commonly extracted from 76 studies. Lau, Zhao and Xiao (2004) found that firms that practice TQM have superior performance in leadership; strategic planning; customer and market focus; measurement, analysis, and knowledge management; human resource focus, process management, and business results.

While a substantial body of literature has been developed linking TQM system to business performance, Kim-Soon and Jantan (2010) reported that there is a dearth of evidence of comparing the soft factors and hard factors and its impact on business performance between big firms and the SMEs. Thiagarajan and Zairi (1997) regarded that systems, tools, and techniques such as quality management systems, cost of quality and statistical process control and external effectiveness (e.g. benchmarking and customer satisfaction surveys) are examples of hard quality factors and the soft quality factors are intangible and are primarily related to leadership and employee involvement. Firms can achieve superior business performance if they spend their resources towards improving their quality management system towards enhancement of their soft as well as hard quality factors. SMES should enhance their quality management to achieve the level of business performance of large firms (Kim-Soon and Jantan 2010).

5. Approaches of implementing TQM

Implementing TQM can be a tedious journey in an organization. Empirical evidence supports the argument that by focusing on quality, a business can substantially improve its performance (Peters and Austin, 1985; Yahya et al., 2001). Different approaches are practiced by organizations to initiate and implement TQM. TQM GOAL/QPC Research Committee, a nonprofit organization located in the United States of America has documented some of these approaches described as follow.

-   The Guru Approach. This method takes the teachings and writings of one of the leading quality thinkers and uses them as a benchmark to determine where the organization has deficiencies and then to begin to make appropriate changes to remedy those deficiencies. For example, managers would attend Dr. W. Edwards Deming's courses and study his "Fourteen Points." They would then go to work on implementing them.
-   The TQM Element Approach. This approach takes key systems, organizations, and tools of TQM and begins work on them. This method was widely used in the early 1980s by companies that tried to implement parts of TQM as they learned them. Examples of this approach included use of specific elements such as Quality Circles, Statistical Process Control, Taguchi Methods, and Quality Function Deployment.
-   The Company Model Approach. In this approach individuals or organizational teams would visit U.S. companies that were taking a leadership role in TQM and determine what successes they had and how they had accomplished them. The individuals or teams would then integrate these ideas with their own and thus develop their own organizational model which would be adapted for their specific organization.
-   The Japanese Total Quality Approach. Organizations utilizing this method take a look at the detailed implementation techniques and strategies employed by Deming Prize-winning companies and use this experience as a way to develop a five-year Master Plan for in-house use.
-   The Prize Criteria Approach. Using this model, an organization uses the criteria of the Deming Prize or the Baldrige Award to identify areas for improvement. TQM implementation under this approach is focused on Prize criteria benchmarks.

There is no one best way to organize quality management system in an organization as it is necessary to fit to the needs of the organization concerned. It is like what Scott (1981) described the contingency approach, "The best way to organize depends on the nature of the environment to which the organization must relate". The business settings are unique, the nature of business itself, the organization cultures and people are different from one another. Thus, the notion of no one right approach of implementation.

6. The quality journey

The journey of quality management never ends. Quality management is evolving and tomorrow will present a different scenario through adding and discarding practices. Whether it is a big organization or a small one, producing products or services, it is quality that matters to the customers. Global market competitiveness and market dynamics are continuously changing the landscape of managing quality. Competitive advantage requires constant corporate attention to the latest definition of customer-driven value. Effective quality management systems are dynamic, adaptable to change in meeting customer's requirements and expectations. It can provide guidance for establishing an organization's processes for maintaining records, improving processes and systems, and meeting customer's requirements and expectations.

Some recent themes in quality management that have become more significant include quality culture, the importance of knowledge management, innovation and the role of leadership in promoting, Kaizen and achieving high quality. Systems thinking are bringing more holistic approaches to quality. A system approach enables an organization to gain and retain customers and to improve overall its efficiency and profitability. Many authors have stressed that operations quality programmes should be both strategic and comprehensive (Slack and Lewis, 2008). They have also prescribed on how TQM integrate into a business strategy (e.g. Lile and Lacob, 2008). TQM has been a popular business strategy in many manufacturing organizations in the past few years (Sohal & Terzivski, 2000). It thus provides evidence of the importance of TQM practices as an effective pillar of corporate strategy for achieving organizational excellence. Thus, for quality management to be strategic, organization needs to commit to an ongoing effort to improve the quality of products, services or processes to sustain market competitiveness of its product and service.

7. Concluding remarks

Total Quality Management is a management philosophy on how to approach the organization of quality improvement through the "holistic" approach. The TQM practices have evolved and improved continually over time to sustain organizational competitiveness. This chapter has dealt on what is quality and TQM, cost of quality, linking quality management system to organizational performance, its impact on organizations and approaches of implementing TQM and the quality journey. For quality management to be strategic, organization needs commit to a continuous improvement journey to sustain market competitiveness of its product and service.

8. References

Ahire, S. L. 1997. Total Quality Management interfaces: An integrative framework. *Management Science,* 27 (6) 91-105.

Anderson, S. W. and Sedatole, K. (1998). Designing quality into products: the use of accounting data in new product development, *Accounting Horizons,* 12(3), September, 213-233.

Anderson, S. W. (1995). A framework for assessing cost management system changes: the case of activity-based costing implementation at General Motors, 1986-1993, *Journal of Management Accounting Research,* Fall, 1-51.

Chong, V. K. and Rundus, M. J. (2004). Total quality management, market competition and organizational performance. *British Accounting Review,* 36, 155-172.

Chowdhury, Subir (2005). *The Ice Cream Maker: An Inspiring Tale About Making Quality The Key Ingredient in Everything You Do,* New York: Doubleday, Random House.

Cooper, R., Kaplan, R.S., Maisel, L. and Oehm, R. (1992). *Implementing Activity-Based Management: Moving from Analysis to Action,* Montvale, NJ: Institute of Management Accountants.

Crosby, Philip (1979). *Quality is Free,* New York: McGraw-Hill.

Cua, K. O., McKone, K. E. and Schroeder, R. G. 2001. Relationships between implementation of TQM, JIT, and TPM and manufacturing performance, *Journal of Operations Management,* 19 (6) 675-694.

Deming, W. E. (1986). *Out of the Crisis*, Cambridge, Mass: Massachusetts Institute of Technology, Center for Advanced Engineering Study.

Drucker, Peter (1985). *Innovation and entrepreneurship*, Harper & Row.

Feigenbaum, A. V. (1991), *Total Quality Control* (3rd ed.), New York, New York: McGraw-Hill, pp. 130–131.

Feigenbaum, A.V. (1956). Total Quality Control. *Harvard Business Review*, 34(6).

GOAL/QPC Research Report No. 90-12-02 (1990),12B Manor Parkway Salem, New Hampshire.

Kaynak, H. (2003). The relationship between total quality management practices and their effects on firm performance, *Journal of Operations Management*, 21, 405- 435.

Kim-Soon, N. & Jantan, M. (2010). Quality Management Practices in Malaysia: Perceived advancement in Quality Management System and Business Performance, *IEEE ICMIT Conference*, Singapore.

Lau, R. S. M., Zhao, X. and Xiao, M. (2004). Assessing quality and management in China with MBNQA criteria. *The International Journal of Quality & Reliability Management*, 21, 699-709.

Lewis, W.G., Pun, K.F. and Lalla, T.R.M. (2006). Exploring soft versus hard factors for TQM implementation in small and medium-sized enterprises. *International Journal of Productivity and Performance Management*, 55(7): 539-554.

Lile, R. and Lacob, M.L. (2008). *Integrating TQM into the Strategy of the Business*, Fascicle of Management and Technological Engineering, Annals of the Oradea University. Vol.VII, (XVII).

Motorola University. "What is Six Sigma?". Motorola, Inc.. http://www.motorola.com/content.jsp?globalObjectId=3088. Retrieved on 01/1/2012.

Peters, T. and Austin, N. (1985). *A passion for excellence*, Collins, Oxford.

Prajogo, D. I. and Sohal, A. S. (2006). The relationship between organizational strategy, total quality management (TQM), and organizational performance-the mediating role of TQM. *European Journal of Operational Research*, 168, 1-20.

Rahman, S. and Bullock, P. (2005). Soft TQM, hard TQM and organizational performance relationships: An empirical investigation, *Omega*, 33, 73-83.

Saraph, J., Benson, P. and Schroeder, R. (1989). An instrument for measuring the critical factors of quality management, *Decision Science Journal*, 20: 810-829.

Scott, W.R. (1981). *Organizations: Rational, Natural, and Open Systems*, Englewood Cliffs NJ: Prentice Hall Inc..

Shah, R. and Ward, P. T. (2003). Lean manufacturing: context, practice bundles, and performance. *Journal of Operations Management*, 21, 129-149.

Shepard, N. (1995). The bridge to continuous improvement, CMA Magazine, March, 29-32.

Smith, M. (1998). Innovation and the great ABM trade-off, *Management Accounting-London*, 7(5(1), Jan, 24-26.

Sila, I. (2005). The influence of contextual variables on TQM practices and TQM organizational performance relationships. *The Business Review, Cambridge*, 4, 206-210.

Sila, I. and Ebrahimpour, M. (2002). An investigation of the total quality management survey based research published between 1989 and 2000. *International Journal of Quality & Reliability Management*, 19(7): 902-970.

Singh, P. J. and Smith, A. J. R. (2004). Relationship between TQM and innovation: An empirical study. *Journal of Manufacturing Technology Management*, 15, 394-401.

Slack, N. and Lewis, M. (2008). *Operations Strategy*, 2nd Edition. Prentice Hall, Financial Times.

Smith, M. (1998). Innovation and the great ABM trade-off, Management Accounting-London, 76(1), Jan, 24-26.

Sohal, A.S. and Terzivski, M. (2000). TQM in Australian manufacturing: factors critical to success. *International Journal of Quality & Reliability Management*, 17 (2), 158-167.

Steimer, T.E. (1990). Activity-based accounting for total quality, *Management Accounting*, October, 39-42.

Taguchi, G. (1992). *Taguchi on Robust Technology Development*, ASME Press.

Thiagarajan, T. Zairi, M. (1997). A review of total quality management in practice: understanding the fundamentals through examples of best practice application – Part III, *The TQM Magazine*, Vol. 9(6), 414-417.

Walton, Mary, and Deming W.E. (1988). *The Deming management method*, Perigee. pp. 88.

Wikipedia, http://en.wikipedia.org/wiki/Quality_(business). Retrieved on 01/1/2012.

Yahya, S., Salleh, L.M. and Keat, G.W. (2001). A Survey of Malaysian Experience in TQM. *Malaysian Management Journal*, 5 (1&2): 89-105.

2

The Criticality of Quality Management in Building Corporate Resilience

Simmy Marwa

Portsmouth Business School, University of Portsmouth, Portsmouth
UK

1. Introduction

The link between the concept of quality and corporate demise has been long established to hold true. In a survey of 120 organizations wound up between 2000 and 2007 in five continents, Marwa and Zairi (2008a) established that there was some causation relationship between quality management and corporate demise. Ignorance of quality, if unchecked, inevitably triggers corporate collapse. Stamping corporate demise thus calls for managers to give best practice quality management principles a chance to thrive than they've been willing to do before. This is particularly crucial, with the economic recession, where corporations are worried about their level of economic exposure and what they should do to survive these troubled times? Bryan and Farrell (2008) argue that companies that acted prudently in the autumn of 2008 to cushion themselves against the adverse effects of recession, may not have probably produced short-term earnings, but may well have underwritten their survival. Arguably, attending to corporate quality flaws must therefore be at the heart of any effective post-recession strategy in order to build corporate resilience.

The article explores the linkage between recession recovery strategies and their association with the concept of quality management. Literature on recession/post-recession recovery strategies is reviewed and constructs of quality management inherent extracted. These constructs are then compared with those obtained from the literature review on quality management to establish commonality, which, if present, should point to the criticality of quality management in any effective post-recession recovery strategy.

Resilience as the on-going ability to anticipate and adapt to critical strategic shifts (Reeves & Daimler (2009), will become an increasingly important driver of future competitive advantage. Nonetheless, strategic resilience is not about responding to a one-off crisis, rather it is about rebounding from a setback (Hamel & Valikangas, 2003). Thus, strategic resilience, which is at the heart of this inquiry, is about continuously anticipating and adjusting to deep secular trends that can permanently impair the earning power of a core business. It is having the capability to change before the case for change becomes painfully clear. Hamel & Valikangas (2003) contend that resilient companies should balance between perceived opportunities and risk, while fostering attributes that prop them up, namely:

a.   *Foresight* – actively monitoring key mega-trends and events that could create opportunities and risks to create a shared internal view of such trends,

b. *Agility* –willingness, preparedness and ability to adapt rapidly in response to, or in anticipation of, disruptive events and ensure a rapid response,

c. *Staying power* – stamina (human/financial) resources to survive unexpected changes,

d. *Entrepreneurialism* – exploring, designing and scaling new business models to fend off rivals,

e. *Diversity* – varied perspectives that are brought to bear on the task of adapting to new environments.

Prudent quality management is essentially about fostering these attributes and, as argued by Leavy (2009) better run companies like Nokia, Microsoft, General Electric and others are taking the need to invest in the future very seriously, even as they clearly consider improving efficiencies and reign on costs. So, it must be recognized by companies that cost control strategies alone will not be sufficient to help them set the pace within respective industries during recovery; and once the momentum behind an emerging business opportunity has dwindled; it is rarely easy to regain. Blieschke and Warner (1992), postulate that the way an entity reacts to signs of economic distress can make or break it, as evidenced in the case of State Bank of South Australia (an entity that had hitherto implemented TQM in 1988). Faced with economic distress of the 1990s the State bank overreacted by eliminating quality initiatives in favour of such short-sighted cures such as; down-sizing, restructuring and programme cuts. Initially the banks' TQM initiative successfully fostered teamwork, employee participation, innovation, risk-taking, and a corporate culture that focused on the customer. However, following recession, the bank lost its leaders, its reputation, and a great deal of money and performance plummeted. It was only when interest in quality was renewed in 1991, in form of a myriad of quality initiatives (cost of quality measurement, focus on waste prevention and keeping close to the customer) that the downturn was reversed. Thus, keys to enduring a biting recession must be an insistence on sustenance of the corporate quality initiatives whatever the cost.

2. Quality management

The concept of quality has been variously defined as 'satisfying customers' requirements continuously' (Kanji, 2001; Marwa and Zairi, 2008b), 'what customers perceive it to be' (Gronroos, 2005) and 'satisfying and delighting the end customer' (Zairi, 2005, Marwa and Zairi, 2009a). The rallying call of the quality concept is customer delight, which must be at the heart of any quality initiative. By extension Total Quality Management (TQM) refers to 'obtaining total quality by involving everyone's daily commitment' (Kanji, 2001; Marwa & Zairi, 2008b), and 'a positive attempt by organizations to improve structural infrastructure, attitudinal, behavioural and methodically ways of delivering to the end customer' (Zairi, 2005; Marwa & Zairi, 2009b). The common themes or constructs of excellence that commonly occur in the literature are summed up in Tab. 1.

In sum, the ten dominant constructs of excellence extracted from the above were:

a. *Leadership* – setting clear directions and values for the organization, creating customer focus and empowering the organization;

b. *Customer focus* – creating sustainable customer value;

c. *Strategic alignment* – emphasis on strategic development, alignment and planning;

d. *People management* – knowledge, skills, creativity and motivation of its people;

e. *Partnership development* – development of strategic long-term mutually beneficial partnerships with external partners (customers, suppliers);

f. *Process management* – effective management of processes that are the "engines", that deliver every organizational value proposition;

g. *Agility*- ability to move and change direction and position of the organization quickly and effectively while under control;

h. *Social responsibility* – encapsulates responsibility to the public, ethical behaviour, and good citizenship;

i. *Communication* – improvement of information flow; openness & transparency within and without;

j. *Continuous learning, innovation & improvement* - challenging the status quo and effecting change by utilizing learning to create innovation and improvement opportunities at individual or corporate level.

The presence of these constructs in a post-recovery recession strategy should therefore confirm the necessity of quality management in any effective post recovery strategy.

	Author	TQM/Excellence Construct	Representative Constructs
1.	Zairi and Whymark (2003) (11 constructs)	- *Leadership* - *Supplier quality management* - *Vision* - *Evaluation* - *Process control & improvement* - *Product design* - *Quality systems improvement* - *Employee participation* - *Recognition & reward* - *Education & training,* - *Customer focus*	
2.	Tari (2005) (7 Constructs)	- *Leadership,* - *Continuous improvement* - *Employee fulfillment* - *Customer focus* - *Process management* - *Quality data reporting* - *Partnership management & public responsibility*	
3.	Lytle (2004) (10 constructs)	- *Leadership* - *Service vision* - *Employee empowerment* - *Customers contacts service technology* - *Service standards and communications* - *Service failure prevention* - *Service failure recovery,* - *Service training & development* - *Service reward and recognition*	1. *Leadership,* 2. *Customer focus,* 3. *Process management* 4. *People management* 5. *Partnerships development* 6. *Social responsibility* 7. *Continuous improvement*

Author	TQM/Excellence Construct	Representative Constructs
4. Dayton (2003) (10 Constructs)	- *People & customer management* - *Supplier partnerships* - *Communication* - *Customer satisfaction* - *Orientation & operational quality* - *Improvement measurement system* - *Corporate quality culture* - *Quality planning*	8. *Strategic alignment* 9. *Agility* 10. *Communication*
5. Xie et al. (1998) (9 Constructs)	- *Leadership* - *Impact on society* - *Resource management* - *Strategy & policy* - *Human resource management* - *Process quality* - *Results* - *Customer management & satisfaction* - *Supplier/Partner management & performance*	
6. Nuland et al. (2003) (9 Constructs)	- *Results orientation* - *Customer focus,* - *Leadership & constancy of purpose* - *Management by process and facts* - *People development & involvement,* - *Continuous learning, innovation and improvement* - *Partnership development* - *Corporate social responsibility*	
7. Hubbard et al. (2007)	- *Effective execution* - *Perfect alignment* - *Adopt rapidly* - *Clear and fuzzy strategy,* - *Leadership* - *Looking out, looking in* - *Right people* - *Manage everything* - *Balance everything*	
8. Porter & Tanner (2004) (8 constructs)	- *Leadership* - *Customer focus* - *Strategic alignment* - *Organizational learning/innovation & improvement* - *People focus* - *Partnership development* - *Results focus* - *Social responsibility*	

Table 1. Constructs of excellence/TQM

3. Corporate resilience

Likewise, a literature review of recession/post-recession recovery strategies rendered it possible to extract constructs of quality management inherent therein (Tab. 2).

Author	Strategy	Elements of Strategy	Quality construct
Berman, Davidson, Longworth & Blitz (2009)	Do more with less	*Cut costs strategically*	*Alignment*
		Conserve working capital	
		Protect cash reserves	
		Increase Flexibility & Responsiveness	*Agility*
	Focus on the core	*Create value for clients*	*Customer focus*
		Strip out non-value add activities	*Customer focus*
		Shift from fixed to variable costs	
	Understand customers	*Target value-oriented customers*	*Customer focus*
		Reduce complexity	*Process management*
	Exploit opportunities	*Disrupt weaker competitors*	
		Focus on growth markets	*Innovation*
		Acquire bargain-price assets	
	Build capabilities	*Protect and acquire critical talents*	*People management,*
		Establish infrastructure	*Process management*
		Invest in innovation	*Innovation*
	Change Industry	*Understand impact of current transformation*	*Process management*
		Pioneer new industry approaches	*Innovation,*
		Cultivate strategic partnerships	*Partnerships,*
	Make speed competitive advantage	*Empower leaders*	*Agility*
		Communicate strategy quickly, clearly and often	*communication*
	Manage risk	*Reduce risk & increase transparency*	*Communications*
Hamed (2009)	HR & L& D can add value	*Map activities & dependencies*	*Process management*
	Retain knowledge if not the person	*Create reservoir of knowledge*	*Learning*
	Communities of best practice	*Share best practices*	*Benchmarking*
	Learning	*Build capabilities*	*People management*
Review (2004)	Training people	*Building capability*	*People management*

Author	Strategy	Elements of Strategy	Quality construct
Review (2009)	Defend or attack,	*Survive, exploit or disrupt,*	*Customer focus*
	Respond to unpredictability	*Monitor change*	*Agility*
		Make decision quickly	*Agility*
		Willingness to change quickly	*Agility*
		Challenge long held beliefs	*Learning*
	Collaborate	*Tap into social networks*	*Partnerships*
		Liaise with struggling suppliers & consumers	*Partnerships*
		Partnerships with consumers	*Partnerships*
Singh, Garg & Sharma (2009)	Lean Strategy	*Cross training of employees*	*People management*
		Customer communication	*Communication*
		Reduced scrap & wastage	*Process management*
		Flexibility in reacting to changes	*Agility*
		Pareto principle (20/80 Rule)	*Process management*
Reeves & Deimler (2009)	Behavioural strategies	*Adjusting the customer offering*	*Customer centricity*
		Exploring new price models	
		Entering new markets & exiting old ones	*Agility*
		Making opportunistic	*Alignment*
	Social strategies	*Partnering with suppliers or customers*	*Partnerships*
		Partnering with competitors	*Partnerships*
		Shaping the environment	*Social Responsibility*
	Reproductive/rege nerative strategies	*Rapid prototyping*	*Innovation*
		Incubation	*Innovation*
	Evolutionary strategies	*Innovation*	*Innovation*
Leavy (2009)	Cost-control strategies		*Alignment*
	Fresh ideas		*Learning*
	Investing in R&D		*Continuous improvement*
	Talent management		*People management*
	Customer care	*Current & potential customers*	*Customer focus*
Birdsall (2007)	Prepare when others are still indoors	*Focused attention to customer preferences*	*Customer focus*
	Give your brand a physical measurement	*Diagnostic analysis of brand*	*Performance measurement*

Author	Strategy	Elements of Strategy	Quality construct
	Focus on your strongest performers,	*Brand performance*	*Performance measurement*
	Teamwork		*People management*
	Teamwork reward		*People management*
	Ongoing brand management		*Innovation*
Selmer & Waldstrom (2007)	Careful downsizing	*Don't fire wrong people*	*People management*
Desmond (2009)	Talk to customers	*Find out their needs*	*Customer focus*
	Improvement & innovation	*Ensure drivers to innovate comes from the top*	*Leadership,*
		Environment to liberate creative thinking	*Social responsibility*
		Innovation systems to review creative ideas	*Innovation*
		Improvements part of every employees objective	*Leadership*
		Improvement focused learning	*Learning*
		Celebrate & reward efforts and success	*People management*
Owen (2009)	Look around you	*Copy idea*	*Agility*
		Solve problem	*Customer focus*
		Experiment ideas	*Innovation*
		Discount model	
Deans, Kansa, Mehltretter (2009).	Offensive strategy	*Revenue growth*	
		Acquisition of weaker competitors	*Alignment*
		Innovation of products or services & processes	*Innovation*
		Strategic investment	*Alignment*
	Defensive strategy	*Protecting revenues*	
		Staying in tune with customers	*Customer focus*
		Protecting profit margins	
		Benchmarking	*Benchmarking*
		Preserving balance sheet assets	
Thomson (2009)	Stay focused		*Alignment*
	Debt management		
	Funding freeze		
	Good times don't last forever	*Advance planning vital*	*Alignment*

Author	Strategy	Elements of Strategy	Quality construct
	Prudence in financial management		
	Cut-throat competition		
	Customer treatment		*Customer centricity*
	Supply chain management		*Partnerships*
	Poor management		*Leadership*
Hamel & Valikangas (2003).	Conquering denial – anticipate challenges	*Management seeing change as it happens*	*Leadership*
		Allow free thinking and criticisms from within	*Innovation*
		Anticipate strategic decays	*Continuous improvement*
		Deal with eviscerated strategies	*Alignment*
		Identify strategies being replicated by competitors	*Agility*
		Identify strategies being supplanted	*Agility*
	Valuing Variety	*Conduct breakthrough experiments year round*	*Innovation*
		Test promising strategies	*Process management*
	Liberating resources	*Free resources to support a broad array of strategy experiments*	*Alignment*
		Don't overinvest on "what is" rather than "what could be"	*Innovation*
		Relocate capital and talents to new initiatives	*Alignment*
		Distinguish between new and risky ideas	*Performance measurement*
		Get funds in the right hands	*Alignment*
	Embracing paradox	*Accelerate pace of strategic evolution*	*Continuous Improvement*
		Lay groundwork for perpetual renewal	*Continuous improvement*
		Enhance operational efficiency	*Process improvement*
	Make sense of environment	*Restless exploration of new strategic options*	*Innovation*
		Realign resources faster than rivals	*Alignment*

Author	Strategy	Elements of Strategy	Quality construct
Jimena (2009)	Focused CSR strategy	*Companies cannot thrive in collapsing societies*	*Corporate responsibility*
		Investing in society by building a whole new customer and suppliers base for future business	*Social responsibility*
Bellingham & Jones (2010)	Differentiation the key to success in the post-recession world	*Values & service – ethics, honesty*	*Customer service*
		Communicate – open and honesty	*Communication*
		Take responsibility – sustainable and collaboration	*Social responsibility*
		Cognition & connectivity – educate and inspire	*People management*
Birkinshaw (2010)	Critical need to reinvent management	*Understand – be explicit about management principles used*	*People management*
		Evaluate – assess suitability management principles to current environment	*Process management*
		Envision & experiment – try new practices to reinforce the choices made	*Innovation*
Crainer (2010)	Create products/services that add value to society	*Value creation*	*Customer focus*
Hubbard, Samuel, Cocks and Heap (2007)	Attributes of winning organizations in Australia	*Effective execution – timely delivery of results*	*Agility*
		Perfect alignment – systems alignment	*Alignment*
		Rapid adoption – rapidly accommodates change	*Agility*
		Clear & fuzzy strategy – mix-up	*Alignment*
		Leadership – build talents	*Leadership*
		Looking out, looking in – SWOT analysis	*Alignment*
		Right people	*People management*
		Manage the downside – be cautious	*Leadership*
		Balance everything – all round focus	

Author	Strategy	Elements of Strategy	Quality construct
Emerton (2010)	Global talent hunt	*Value – people with appropriate values*	*People management*
		Development – grow talent	*People management*
		Talent management –core strategic capability	*People management*
Gulati, Nohria and Wohlgezogen (2010)	Responses to slow down	*Prevention focus – makes primarily defensive moves*	*Process management*
		Promotion focus – invest more on offensive moves	*Agility*
		Pragmatic – combines both defensive and offensive	*Agility*
		Progressive – deploy optimal combination of defense and offense	*Agility*
Rao (2010)	Attracting employees	*Justice – values and rules that bind everyone*	*People management*
		Transparency – open to employees	*Communication*
		Learning – growing talents	*People management*
		Competence	*Innovation*
		Fun – work life balance	*People management*
		Flexibility – is work flexible?	*People management*
	Attracting suppliers	*Vendor relationship – antagonistic?*	*Partnership*
	Attracting shareholders	*Transparency and authenticity*	*Communication*
	Attracting consumers	*Delivering value*	*Customer focus*
Bird, Flees & DiPaola (2010)	Talent gap	*Quantify your leadership gap – rigorous analysis of leadership gap*	*Leadership*
		Deploy current talent more effectively -	*People management*
		Reduce demand for talent – use what is available effectively	*People management*

Table 2. Post recession recovery strategies

A closer examination of the constructs extracted (Tab. 2) revealed some commonality or semblance with those attributes of quality management (Tab. 1), suggesting that excellence should be an integral component of any effective post-recession recovery strategy (Tab. 3).

	Table 2 - Quality Constructs	Frequency (No. of authors)	Table 1 – Quality Constructs	Differences
1.	Agility	6	*Agility*	
2.	Customer focus	12	*Customer focus*	
3.	People management	6	*People management*	
4.	Leadership	3	*Leadership*	
5.	Innovation/ Continuous improvement/ Learning	7	*Innovation/Continuous improvement/Learning*	
6.	Partnerships	4	*Partnership development*	
7.	Communication	4	*Communication*	
8.	Alignment	3	*Strategic alignment*	
9.	Benchmarking	2		Benchmarking
10.	Work environment	2		Work environment
11.	Performance measurement	2		Performance measurement
12.	Process management		*Process management*	

Table 3. Elements of quality management

Of the twelve constructs of excellence extracted from the strategies, nine corresponded (word for word) with those originally identified as constructs of excellence (Tab.1). Nonetheless, further examination of the three made it possible to conclude that perhaps only one of them, (benchmarking) stood distinct. The other two (work environment and performance measurement) were undeniably embedded in "people management" and "process management"; constructs, hence, already included in the comparative analysis.

Moreover, this exploration revealed that in reality, constructs of quality management/excellence though critical to resilience, rarely feature prominently and independently as standalone recession combat strategies. Rather, often these are lumped together with others and/or concealed beneath other strategies, which thus relegates them to the periphery. The danger with this "concealment" is that quality management issues may not get the attention deserved despite their centrality in building up corporate resilience. The upshot of relegating quality management to the periphery will, if unchecked, frustrate recovery and/or herald corporate instability and ultimate demise.

4. Fortifying resilience via TQM

The old saying that "the best defence is a well thought-out offence" applies both in times of recession too. So, the best business approach for combating recession is to plan for recession ahead of time. A management team that is adroit at planning and implementing recession strategies may actually use the circumstances of recession to expand its market share (Bigelow and Chan, 1992). There has to be a shift in focus from short-termism objectives to long-term survival and health of the enterprise. Corporations must be willing to invest more resources than ever in total quality analysis, just to look for leverage points and errors to improve upon (Vargo and McDonough, 1993). There has to be genuine concerted efforts to continually build up corporate resilience, not on short-term basis but rather enduring longer objectives. Industries must focus on maintaining portfolio quality and creating effective risk management strategies (O'hare, 2002). Rafts of measures have been suggested to prop-up corporations and stop them from sliding into demise.

Suffice to argue that even in the face of the current global economic meltdown, companies must continue to pursue quality and pay the price, because the alternative is unpalatable– demise! It is with these clear options in mind that building a robust post-recession recovery strategy must essentially embrace elements of quality management as listed hereunder:

4.1 Leadership

Leaders must become aware of their business practices by re-examining everything a company does and making emotional connection with its management team. Leaders have to realize that no one person can make a company successful during a period of recession; it requires more hands. However, one person with a command of leadership amongst the management team can transfer sufficient influence by creating an enabling environment for the management group to thrive and guarantee success. Bigelow and Chan (1992) assert that leaders should plan for recessions or downturns the same way they do in good economic times. Executives must now take a more flexible approach to planning: each of them should develop several coherent, multi-pronged strategic action plans, not just one and, pursue them quickly as the future unfolds (Bryan and Farrell, 2008). Managers who find themselves in a recession without pre-planned strategies must now spend time analysing situations carefully, identifying predominant issues affecting their businesses and carefully developing appropriate goals and strategies, bearing in mind that post-recession a position is just as important, if not more, important than immediate cost cutting (O'Hare, 2002). During a recession, progressive leaders encourage their organizations to discover what works and combine those findings in a portfolio of initiatives that improve efficiency along with market and asset development (Gulati, et al., 2010). Smart leaders must insist on scoring decisions, so as to utilize resources effectively, reduce costs and create better customer relations that positively impact the bottom-line.

4.2 Corporate responsibility

WBCSD (2008) reports that a number of companies are gradually pursuing responsible CRS strategy and cite Wal-Mart, which has just signed a sourcing agreement with its suppliers that require factories to certify compliance with certain environmental standards and laws, which are now being applied to filter out suppliers. When leaders exercise discipline and focus on mobilizing employees to respond to customers' interests and values, they increase the chance that, when the downturn ends, they will come out on top (Shirazi, 2008). Tan (2008) and Thawani (2008) reinforces the view that in a downturn, business improvement methodologies provide a more prudent approach for reviewing organizations and should be utilized by companies that are seeking to reinvent themselves in terms of practice, processes and human resources. WBCSD (2008) has reported that organizations that instituted effective management and supplier management systems, reduced production wastes, improved planning and business forecasting, have continued to thrive, even in the face of a recession.

4.3 People management

People management relates to both internal and external customers. Leaders who understand best practice and invest in their employees, through development and training will maintain competitive advantage and grow market share by taking it from the competition (Johnston, 2008). Shirazi (2008) asserts that since high-impact talents tend to be poached by competitors in a downturn, companies should provide development experiences and rotational assignments that ensure high retention. Retraining and developing top talents ensures that everyone is informed and encouraging questions, new ideas, creativity and innovations (Coombe, 2008). Leaders must therefore, encourage constructive conflicts that challenge the status quo and fuel good decision making, by encouraging questions and new ideas, while making it safe for employees to raise them. Management must be open and acknowledge that they don't have all the answers to the myriad of issues and should therefore solicit employee inputs, thereby empowering their people to contribute their best ideas (Shirazi, 2008). WBCSD (2008) has established that companies that have invested in their social capital (improving their manpower skills) and had a transparent rewarding system, have been able to improve their bottom-line without compromising on their social performance and compliance to quality standards during a recession. HRfocus (2010) further contends that talent management must mesh with business goals for post recession success. Companies need to accommodate different lifestyles and work choices and find ways to balance these with business needs to ensure high level of product and performance. The best and most talented people are not going to want to hang round a company that isn't signalling that it has faith in its own future growth potential (Leavy, 2010).

4.4 Customer focus

McKay (2008) contends that businesses can survive the downturn if only they cherish their existing customers. Since it costs 5 to 7 times as much to acquire a customer as it does to retain one, companies should optimize existing relationships by making their customers more appreciated. For example by using the Ritz Carlton formula; treating every customer as if they are their best customers, every business can thrive even in a recession. Such initiatives as respecting customer's time, keeping promises, keeping customer in the information loop, delivering same day customer purchases and showing genuine interest in the customers' satisfaction and success amongst others will pay dividends (Berman, et al., 2009; Reeves & Deimler, 2009). Organizations should always look for additional touch points, while at the same time, anticipating customers' needs and working flat out to meet such needs (Deans et al., 2009). Moreover, since customer service is often a primary target for cost cutting, when hard times hit, it must be kept in perspective, that customers are the reason firms exist. IIf customer service can be re-organized, so that same or superior level of service is offered at reduced cost, then that would be acceptable (Belingham & Jones, 2010; Craner, 2010). However, should cost-cutting measures cause reduced customer service, then that may possibly trigger a chain reaction for the worse from customers to the detriment of organizations. Thus, re-familiarization with own customers, and giving them the best service ever, businesses will guarantee their fastest turnaround possible.

Likewise, Coorey (2008) suggests that beating recession calls for a raft of strategies; significantly knowing why customers buy from an enterprise (define your offer to ensure clarity amongst employees) and ensuring customer satisfaction (on all product/service offerings traded), thereby invigorating the customers' experiences. Choudhury (2008) also argues that CRM is critical during a period of economic downturn and, suggest that companies should work towards improving relationships with existing customers, by investing carefully in ideas and systems that helps keep the customer base connected and reassured, as well as improving their own branding (cleaning-up and positioning brands in the marketplace. Coombe (2008) buys into the view and opines that recession can be effectively overcome through looking after one's customers and, suggests that businesses should work out what they can do for their customers that gives them a better return and extra value for money. Leavy (2009) advocates taking a fresh look at customers and potential customers using two tools: customer consumer chain and a mapping of the attributes of products/services. Consumer consumption chain analysis involves analysis of a whole set of activities that consumers are going through as they seek to meet their own needs, identifying those that are useful and/or might be valuable to the organization, while discarding those that are not useful. Mapping of product/service attributes on the other hand allows for reconfiguring of an organization's offers to better match what customers are really looking for. The mapping takes a customer segment and assesses the extent to which certain attributes are positive, negative or neutral from their perspective, and then assesses their emotional response to these attributes. Sometimes customers may be happy to reward companies for adding more of those things they want and will pay for, or even taking out those which they don't easily care about. Likewise, Reeves and Daimler (2009) support the contention of adjusting the customer offering and argues that a recession may offer opportunities to introduce new services that enhance a firm's proposition for budget strapped customers (new sizes, new looks, price models).

4.5 Agility/process management

Watterson & Seeliger (2008) argue that entities can prosper even in a recession if only they can guarantee their survival, by aggressively investing in operational excellence tools (including Lean and Six-Sigma) - as is evident in some aircraft companies, where staff has been empowered with the right tools to tackle business challenges. Choudhury (2008) equally emphasizes that companies must get lean and efficient, rather than merely cutting costs and jobs. Organizations should consider mapping all processes and embrace workflow efficiencies, through interventions, like Six-Sigma. Similarly, Johnson (2008) concurs that businesses should constantly build advantages into the organization at a much greater rate, while eliminating disadvantages. By taking advantage of opportunities presented by the economic meltdown, organizations should vigorously pursue creativity and innovation, so as to offer their clientele, innovative solutions to current challenges. In addition to simplification of value propositions (highly customizable product ranges), organizations should focus on providing both superior frills/options (for the high end market) and incremental frills/options (for the most basic customer needs). A focus on scientifically determined features that drive customer choice and purchase, and consequent trimming down of product bundle to what can be offered profitably, will enable entities to maintain a firm grip on their markets (Singh et al., 2009).

Proactive entities should embrace recession because of the opportunities it offers (Gilati, et al., 2010). By being the first to anticipate the recession's impact and acting quickly to provide customized solutions that leads to dramatic improvement in relative positioning and performance will secure such entities a firm footing. Sewell (2008) challenges organizations to prioritize activities by auditing resources they're using to deliver services and provide clarity on all services being provided. In sum, it's about corporations making sure that their enterprise performance management engine is properly attuned to support the entity in these exciting times (The Hackett Group, 2008).

4.6 Learning/innovation & improvement

In conditions of high uncertainty, organizations need to invest in learning rather than in analysis. Leaders have to take step-by-step approach to innovation and growth that limits risk exposures at each step, while maximizing their learning capacity (Leavy, 2009). Since recession forces customers to make tradeoffs (Reeves & Daimler, 2009), organizations should be on the look-out for new opportunities for growth; exit or entry into new markets, different potential offerings and/or vital changes to business operations that requires adoption to reflect the changing consumer demand (Rao, 2010). A crisis is a chance to break ingrained structures and behaviours that sap the productivity and effectiveness of many organizations (Bryan and Farrell, 2008). Such moves aren't short-term crisis-response, but rather long-term.

Therefore smart leaders must know their corporations well during crisis and demonstrate this knowledge in managing challenges arising from illiquidity, plummeting profits, employment and redundancies (Coombe, 2008). The strains resulting from the crisis generated by the recessions must continually be reviewed and flexible approaches adopted that cushions the entity against further stress. Roles need to be streamlined to offer greater collaborations and prevent antagonistic and wasteful competition (Desmond, 2009). During difficult times, employees can either be the most valuable asset or the biggest liability (McGhee, 2008). The difference between those two outcomes depends on how employees are handled.

4.7 Communication

Good work environment and communication between an organization and its employees is critical and more so times of recession, where the team approach should be embraced in all forms of communication (Bellingham & Jones, 2010). Business leaders should keep their employees informed, engage with all staff inclusive of those at lower level ranks and make sure everybody in the organization knows how the organization is faring. Since staff will always perceive if something is amiss, being open with them in difficult times often pays dividends in terms of loyalty and commitment (Coombe, 2008). Shirazi (2008) suggests that business leaders should seek to manage the heart, communicate authentically (honestly, freely and openly or transparently), create positive vision and attitude to their employees, by acknowledging reality and the challenges they are struggling with, rather than stonewalling them. Leaders are often tempted during difficult times to make unilateral decisions, thereby developing dictatorial tendencies, which is unwise. Authentic brands are more than promises made to employees and staff (Buckingham, 2008); they are promises delivered and there is nothing quite like tough economic conditions to prohibit entities from forging a sharpened customer-focus. Whether faced with a market downturn or not, the internal communication community has a pivotal role to play in ensuring that employees engage with the brand (Ahmed, 2009). To bridge the engagement gap leaders should: always role model an open-door policy; especially in turbulent times, be honest with its people and readily empathize, which are the personal qualities needed in tough times (Rao, 2010). Thus, it is important that leaders take a regular reading of the "corporate temperature" (measure the impact of internal communication constantly) and seek out and promote positive role models and good stories that continually reinforces corporate core values.

4.8 Partnerships/social strategies

Reeves & Daimler (2009) assert that firms should explore whether there are opportunities for banding together "socially" through cooperative relationships:

a. Partnering with suppliers or customers – explore ways of working together to remove direct, inventory, or transaction costs from the system for mutual benefits,
b. Partnering with competitors – explore opportunities of coming together with competitors to say share: services, capabilities or assets. See whether open –source innovations models can lower costs and spur growth,
c. Shaping the environment – is there an opportunity to argue for favourable changes to regulatory regimes – independently or in concert with competitors?

Such partnerships create invaluable networks that companies can rely on to build their excellence programmes and deliver value to their customers.

4.9 Benchmarking

Choudhury (2008) suggests that entities should never stop learning; instead they should have continually access available knowledge base (from best practice entities), rethink corporate staff development strategies and apply them. Cutting wastes and harvesting best practices from other corporations and using them must be the "modus operandi" in today's corporations (Adam, 2008). Corporations should learn from best practice, competitors and be open; the more an entity is open, the more it stands to benefit from a benchmarking initiative (Coombe, 2008). Cavanaugh (1995) observe that in the mild recession of early 1990s, companies that outgrew recession were those that focused on both best practice benchmarking and technical benchmarking (numerical). Whether it is customer focus, relative value, market dominance or indeed, looking for end-to-end, companies who deployed these approaches to improve performance, expanded and/or indeed improved their global standing. Likewise Vargo and McDonough (1993) argue that by seeking out the best practices of other companies, can serve as an eye-opener for an entity in areas with potential for improvement to zero in on. This way, an organization will be able to provide the best value for money and should be the driver that catapults enterprises to success.

5. Implications

That some corporations are struggling to fight off recession is not anything surprising if the evidence adduced in this paper is anything to go by. For some organizations perhaps the concept of quality management, may not have been brought to the fore as an effective means to combat recession, but rather "concealed" beneath other strategies whose effect is relegation of quality management initiatives to the periphery. Leaders need to prioritise quality initiatives and align quality management with their corporate strategies so as robust quality management initiatives are pursued alongside other equally deserving strategic initiatives to prop-up organizations. An effective realignment of quality management with corporate strategy that affords an overlap of initiatives has the potential to underwrite resilience and prop-up organizations in these difficult economic times.

6. Conclusion

In a recession companies are being "squeezed out" if they are not making smarter decisions for these challenging times and must therefore dominate through superior quality management, smarter decision making and rapid responses to challenges through overall improvement of efficiency. In times of economic hardship smart corporations not only survive' but also position themselves to take advantage of opportunities that arise as the market improves. Top performing corporations continually invest in their people, processes and technology by differentiating themselves from the rest of the pack and so should be with all corporations aspiring to survive this recession. Ultimately, the margin between winners and losers in this game will be the quality of their deliverables/offerings. Quality management should be at the centre stage of any effective post recovery strategy and robust quality management initiatives should be pursued as an integral element of a robust recession recovery response strategies to secure corporate sustainability.

7. References

Adams, D. (2008), "A B2B recession survival kit: Three not-so painful tips for thriving in a miserable economy", *Progressive Distributor*, Web page, http://www.progressive distributor.com – (Accessed on 30/12/2008).

Ahmed, A. (2009), "Creating opportunity from crisis: taking a strategic and learning–focused perspective", Development and Learning in Organizations, Vol. 23 (5), pp. 4-6.

Bellingham, M., and Jones, N. (2010), "Differentiation the key to success in the post-recession world", Web page, www.nzbusiness.co.nz – Accessed on 12/4/10.

Berman, S., Davidson, S., Longworth, S., and Blitz, A. (2009), "Succeeding in the economic environment – three targets for leaders", Strategy & Leadership, Vol. 37 (4), pp. 13-22.

Bigelow, R., and Chan, S. P. (1992), "Managing in difficult times: Lessons from the most recent recession", *Management Decision*, Vol. 30 (8), pp.34-41.

Bird, A., Flees, L., and DiPaola, P. (2010), "Start filling your talent gap –now", Business Strategy Review, Spring, pp.57-63.

Birdsall, C. (2007), "Keeping your brand fit when times are tough", A handbook of Business Strategy, Emerald Backfiles, pp. 356-361

Birkinshaw, J. (2010), "The critical need to reinvent management", Business Strategy Review, Spring, pp. 5-11.

Blieschke, E.A., and Warner, G.J. (1992), "Quality the tool to survival", *Annual Quality Congress*, Nashville TN, Vol.46 (0), May, pp. 1051-1056.

Bryan, L., and Farrell, D. (2008), "Leading through uncertainty", *The McKinsey Quarterly*, December.

Buckingham, I. (2008), "Communicating in a recession", *Strategic Communication Management*, Vol. 12(3), pp. 7-8.

Cavanaugh, T. (1995), "Quality strategies '95: Benchmarking goes global, *Chemical Marketing Reporter*, Vol. 247 (15), pp. 2-5.

Choudhury, R. (2008), "Being creative: Or how to ride out the global recession". Web page, http://www.nasscom.in – (Accessed on 22/12/2008).

Coombe, S. (2008), "Surviving the recession: ideas and tips to get creative companies through the worst of it", Web page, http://www.smith&william.co.uk – (Accessed on 25/12/2008).

Coorey, D (2008), "Beating the recession", Web page, http://www.37point5.co.uk – (Accessed on 22/12/2008).

Crainer, S. (2010), "Innovating globally", Business Strategy Review, Spring, pp. 25-27.

Deans, G.K., Kansa, I.C., and Mehltretter, S. (2009), "Making a key decision in a downturn: go on the offensive or be defensive", Ivey Business Journal, Vol. 73 (1), January-February, pp.3-13.

Desmond, V. (2009), "How a Systematic approach to innovation can bring radical improvements to organizations by liberating staff", Industrial and Commercial Training, Vol.41 (6), pp.321-325.

Emerton, R. (2010), "The global talent hunt", *Business Strategy Review*, Spring, pp. 79-83.

Gubbins, E. (2008), "Surviving the Recession: Business Services", *TelephonyOnline*, Web page, http://www.printthis.clickability.com – (Accessed on 24/12/2008).

Gulati, R., Nohria, N., and Wohlgezogen, F. (2010), "Roaring out of recession", *Harvard Business Review*, March, pp. 63-69.

Hamel, G., and Valikangas, L. (2003), "The Quest for Resilience", *Harvard Business Review*, September, pp. 52- 63.

HRfocus (2010), "HRfocus news in brief", Web page, www.hrfocus.com – Accessed on 12th April 2010.

Hubbard, G., Samuel, D., Cocks, G., and Heap, S. (2007), The First XI: Winning organizations in Australia, John Wiley & Sons Australia Ltd, pp. 16-17.

Johnson, R. (2008), "Creating success during a downturn in the economy", *Supply House Times*, Vol.51 (2), pp.184-188.

Kanji, G. (2001), Total Quality Management: Myths or Miracles, Kingsham Press, Chichester.

Leavy, B. (2009), "Surviving the recession and thriving beyond it: Rita McGrath explains how to deliver discovery-driven growth, *Strategy & Leadership*, Vol. 37(4), pp.5-12.

Lytle, R.S. (2004), "10 elements of service excellence", *Texas Banking*, Vol. 93(6), pp. 22-27.

Marwa, S.M., and M. Zairi (2008a), "Towards an Integrated National Quality Award in Kenya", *The TQM Journal*, Vol. 20 (3), pp.249 -264.

Marwa, S.M., and Zairi, M. (2008b), "An exploratory study of the reasons for the collapse of contemporary companies and their link with the concept of quality", *Management Decision*, Vol.46 (9), pp.1342-1370.

Marwa, S.M., Keoy, K.H., and Hoh, S.C.L. (2009a), "Is Excellence an Elusive Dream for East African Business Schools", *International Journal for Enterprise Network and Management*, Vol. 3 (2), pp.93-111.

Marwa, S.M., Rand, G.K., Keoy, K.H., and Koh, S.C.L. (2009b), "Sustaining Service Excellence, in the Insurance Industry: a Kenyan Perspective", *International Journal of Value Chain Management*, Vol. 3(3), pp.288-301.

McKay, C. (2008), "Cherish your existing customer: surviving the recession", Web page, http://www.allbusiness.com – (Accessed on 25/12/2008).

O'Hare, E. (2002), "Smarter Decisions for challenging times", *Business Credit*, Vol. 104 (3), pp. 54-59.

Owen, J. (2009), "Look around you", Web page, www.director.co.uk – Accessed on 2/02/2010.

Porter, L.J., and Tanner, S.J. (2004), Assessing Business Excellence, Elservier Butterworth Heinemann, pp.6-7

Rao, S.S. (2010), "Is your company ready to succeed?, *Business Strategy Review*, Spring, pp.69-73

Reeves, M., and Deimler, M.S. (2009), "Strategies for winning in the current and post – recession environment", *Strategy & Leadership*, Vol. 37 (6), pp. 10-17.

Review (2004), "Training people – because you are worth it: How top companies survived recession by refusing to cut back on educating staff", Development & Learning in Organizations, Vol. 18 (4), pp. 18-21,

Review (2009), "Looking forward in recession: turning threat into opportunity", Strategic Direction, Vol. 25 (10), pp. 12-25.

Sewell, J (2008), "Surviving Recession: Action Steps", Web page, http://www.tamarack-uk.com - (Accessed on 25/12/2008).

Singh, B. Grag, S.K. and Sharma, S.K. (2009), "Can lean be a survival strategy during recessionary times", *International Journal of Productivity and Performance Measurement*, Vol. 58(80, pp. 803-808.

Shirazi, F.A. (2008), "10 Actions to ride out of recession", Web page, http://www.chiefexecutive.net – Accessed 25/12/2008).

Tan, A. (2008), "Surviving an Economic Recession – A business Improvement Perspective", Web page, http://www.ezinearticles.com - (Accessed on 25/11/2008).

Tari, J.J. (2005), "Components of successful total quality management", *The TQM Magazine*, Vol. 17 No.2, pp.182-94.

Thawani, S. (2009), "Time ripe for businesses to focus on quality management", *Emirates Business*, 5th January.

Thomson, J. (2009), "25 Corporate collapses – and the lessons learnt", Web page, www.smartcompany.com.au/strategy/20090212-25-corporate-collapses-and-the-lessons-learnt - Accessed on 10/02/2010

Vargo, R. P., and McDonough, S.G. (1993). "How to do more with less", *Financial Executive*, Vol. 9(2), pp.41-45.

Watterson, A., and Seeliger, J. (2008), "You can prosper in the next recession", Web page, http://www.Oliverwyman.com - (Accessed on 25/12/2008).

World Business Council for Sustainable Development (WBCSD) (2008), "Brands stay committed despite tough trading", Web page, http://www.wbcsd.org – (Accessed on 25/12/ 2008).

Xie, M., Tan, K.C., Puay, S.H. and Goh, T.N. (1998), "A comparative study of nine national quality awards", *The TQM Magazine*, Vol.10 (1), pp. 30-39.

Zairi, M., and Whymark, J. (2003) Best Practice Organizational Excellence, E-TQM College Publishing House, Bradford.
Zairi, M (2005), Total Quality Management: Gift to the World, Spire City Publishing, Clitheroe, Lancs.

3

The Strategic Approach of Total Quality and Their Effects on the Public Organization

Luminita Gabriela Popescu

National School of Political Studies and Public Administration
Romania

1. Introduction

The main objective of this chapter is to emphasize the main effects produced, in an organization, by the integration of total quality strategy within the general strategy of the organization. For this integration to be possible it is mandatory to construct an adequate context. In the second part of the chapter, the transfer achieved through the integration of quality strategy within the organization's strategy is undertaken.

The research methodology that consists of the institutional analysis (Meyer and Rowan, 1977; DiMaggio and Powell, 1983; North, 1990) is a type of social science research that seeks to reveal the effects on formal policies, informal norms and interpretations of the practices associated representative of a particular area on the actual scope.

The organizations analysis highlights ways in which different social institutions - legal norms, regulations, procedures and their associated meanings - and propose working material and symbolic incentives (reputation, trust, accreditation) and thereby generate configurations of processes, strategies and practices at the organizational level.

Carrying on, we will present the main steps contained by the two parts of the chapter

2. The constitutive context of quality strategic approach

Under the pressure of the citizens whose exigencies are becoming more and more sophisticated; of the new management approaches that imply, for example, the abandonment of the Management by Objective and the endorsement of the Management by Results; of domestic political forces and due to constrains from the global social-political forces, the need for reform and innovation in the public sector is more and more obvious.

In the bureaucratic hierarchy, activities take place according to general rules and norms. The main objective of the structures and the managerial control instruments is ensuring the conformity with the prior rules and norms. In such a system, the improvement of the efficiency and the effectiveness of the processes involve endorsement of several legislative alterations.

2.1 Meta-organization

In this new context "quality approach in public sector is not only a technical measurement and implementation issue. It is also a political problem where changes in quality are connected to government activity and, in the end, to society activity"[1]

This implies that public organizations evolve[2]:

a. from a closed, self-centered service provider to an open networking organization which public trust in society through transparent process and accountability and trough democratic dialogue;
b. from an internal (resources and activities) focus to external (output and outcome) focus;
c. from a classical design-decision-production-evaluation cycle to an involvement of stakeholders in general, and citizen (as customers) in particular *at each and every stage*.

Achievement means giving up old paradigms and acceptance of some innovative approaches in which services beneficiaries / users are, at the same time, co-participants in the innovation of the public service they benefit from. In other words, the development of a new type of relationship between public services providers and their beneficiaries / users is necessary.

Moreover, the new managerial approaches related to quality strategic approach (Management Based on Results) impose closer attention paid to results.

Focusing on results expresses the need for the creation of a strategic vision of the expected finality, vision which exceeds the orders of the organization and which takes into consideration, on the one hand the fructification of the positive influences from external factors, and on the other hand reduction (elimination) of threats coming from them.

Consequently, the innovation of public services according to these coordinates becomes possible only when a *meta-organization* which the organization of public services, beneficiaries / users of public service interested in outputs and other categories of stakeholders interested especially in results be part of, can be achieved The meta-organization, as a flexible network-type structure, is built in such a way that it contains both the organization that provides the service and the actors that surround it. The latter are either customers/users of the service or representatives of the community's interests. In figure 1 the meta-organization is represented for a particular case of a social service.

The actors involved are users/customers of the provided public service, representatives of the community's interests observing the exigencies necessary to the provision of the service and other categories of stakeholders, as well.

We think that the presence of International/European Bodies in the structure of a meta-organization is a gain because the former can be used as "bases of best practice", useful for improving the performances of the provided service.

---

[1] C. , Politt and G., Bouckaert, *Quality Improvement in European Public Services*, SAGE Publication Ltd. London, 1995. p.12.
[2] C., Pollitt, G.,.Bouckaert, E.,Loffler, *Making Quality Sustainable: Co-Design, Co-Decide, Co-Produce, and Co-Evaluate,* The 4QC Conference, Tampere, 2006., pp.5-6.

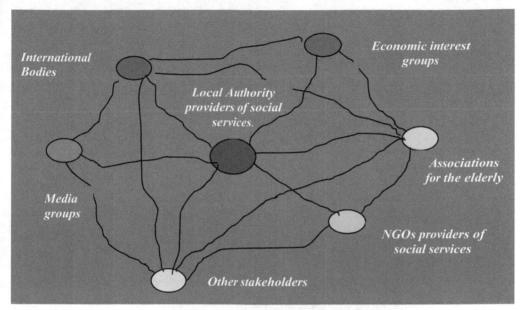

Fig. 1. The Model of Meta-organization- The case of the public service for social protection at local level (Adapted after L.G.Popescu, (2005), Public Policy, Ed.Economica, Bucharest, p.135)

This construction represents a potential *solution based on co-operation* between all the actors of the meta-organization, aimed at creating innovatory public services.

Cooperative solutions are required, not only in the form of co-operation between governments but also through co-operation between governments (centrally, regionally, locally), civil society association and other stakeholders such as media and business (C., Pollitt, G.,.Bouckaert, E., Loffler, 2006, p.8).

As a first conclusion, the configuration of the constitutive context of public services innovation through the strategic approach to quality implies the need for a new strategic and innovative thinking in the relationship between the central administration and the local and regional administrative organizations, between administrations and citizens belonging to local and regional communities, between administrations and different groups of stakeholders.

Secondly, there is an urgent demand to make the central and local administration structures more efficient (for them to become compatible with the flexible structure of the meta-organization) and to limit the decision-making capacity of the administrations by involving citizens and interest groups representative for the community in the decision-making process.

Pragmatically, the achievement of such a structure implies overcoming a variety of challenges. On the one hand, are the members of the community aware of the importance of their commitment? Are they truly motivated to take part in such a structure? On the other hand, how prepared are political representatives and public authorities to accept co-operation with different categories of stakeholders?

First of all, the lack of a *quality culture* with all the actors of the meta-organization (respectively the quality culture of the members of the community) is one of the major difficulties to overcome in reaching the success of this construction. The responsibility of both political and public authorities to enable this structure to become functional must be focused on the development of this type of community culture. Only when community members become aware of the benefits of the innovation of public services through quality and are willing to commit themselves in different forms, the meta-organization will be substantial.

In conclusion, achieving the quality strategy in public services as seen by this paper is impossible without an informed and active community truly involved in the innovation of public services. Mutually, the members of the community cannot reach the level of quality culture that implies commitment and attendance if the responsible agents at the central, regional or local level do not focus their efforts towards both stimulating the members of the community to commit themselves to innovating public services and revealing the advantages of *"listening to the customer's voice"* rather than *"listening to the hierarchy voice"* *(L.G. Popescu 2002, p.56).*

2.2 A new projection of the quality circle

In these circumstances the traditional, purely legal relationship between consumer and provider is replaced by a creative cooperation and collaboration between the actors of the meta-organization.

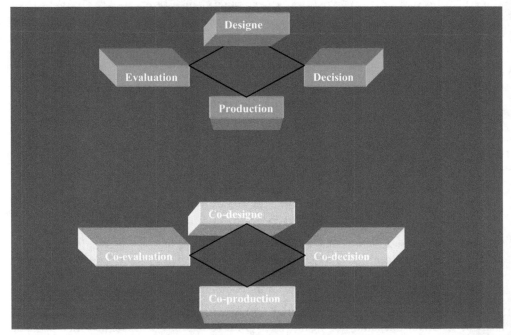

Fig. 2. The shift to co-design, co-decision, co-production and co-evaluation
Source: C., Pollitt, G., Bouckaert, E.,Loffler, (2006), Making Quality Sustainable: Co- Design, Co-Decide, Co-Produce, and Co-Evaluate, Report of The 4QC Conference, Tampere, p. 7.

Moreover, inside the meta-organization, the contradictions between the concepts "consumer" and "provider" and the cooperation and creative dialogue relationships between actors should be revealed.

In figure 2, one can see the result of this deep change determined by the principles on which the new type of relationship develops; from the traditional type where the consumer was "stopped at the gate of the organization" to the new one, where he becomes *co-participant* throughout the quality cycle: *co-design, co-decision, co-produce and co-evaluation* (C., Pollitt, G., Bouckaert, E., Loffler, 2006, p.7).

3. Quality strategic approach

From the meta-organizational point of view, the Strategic Approach to Quality is formulated based on a macro-vision of itself and it is *the ability to simultaneously conduct changes of the consisting systems.*

According to our theory, which we will try to demonstrate here, the Strategic Approach to Quality materializes in a complex change oriented towards four dimensions: *re-defining the potential meta-organization, re-structuring the meta-organization, re-vitalizing the portfolio of the meta- organization and reinventing the mentalities (see fig.3).*

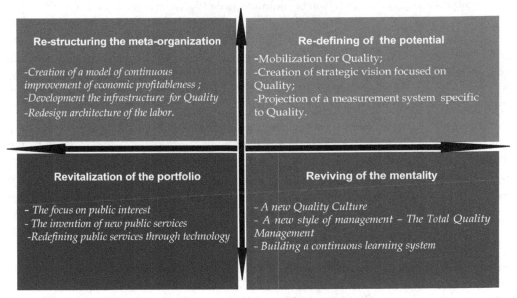

Fig. 3. The stages of Quality strategic approach in public services

3.1 Re-defining the potential of the meta-organization

This dimension implies giving up old patterns and restricting thinking concepts, and accepting new patterns and visions of what the provided public service is and what it could become.

The contents of this transformation include: achievement of mobilization for Quality, creation of a strategic vision focused on Quality and projection of a measurement system for

Quality. In the centre of this transformation there is the development of the energies necessary to the innovation of the service, in other words the mobilization of intelligence and motivation both at the level of the external actors and the internal ones in order to innovate the public service.

3.1.1 Mobilization for quality

Analyses and studies conducted in several organizations in the Romanian public sector have led us to the conclusion that the difficulties occurred during the application of the Quality strategic approach are mainly results of the maintenance of the traditional system where the relationships are essentially bureaucratic and hierarchical.

The political leaders determine what service has to be provided, on what terms and to whom, and bureaucrats and professionals subsequently organize and deliver that service. The role of the citizens is mainly passive. In such a system, completely inadequate for the present-day needs, activities take part without paying attention to the effects on the community / society as a whole. The conditions or quality mobilization to become operational are:

- *The development of an open communication system* mainly characterized by the free circulation of formal and informal messages within the meta-organization. Messages are sent in every direction: on the vertical, in both ways and on the horizontal, without any constrains or inhibitions imposed by the chain of command of the hierarchy in the organization;
- *Participation, meaning the degree in which the internal and external actors involve themselves in the cycle of the quality: design, decision, production, and evaluation.*

Applying these tenets requires managerial efforts oriented firstly towards acknowledgement of the importance of the relationship between the actors and secondly, towards the provision of resources and information, support and freedom of initiative necessary to develop this relationship satisfactorily.

Contrary, there will be negative effects both on the external actors whose requirements have not been met, and on the internal factors which will be frustrated and de-motivated by the lack of finality in their actions.

In conclusion, orientation towards the external factors is reflected through the managerial availability to satisfy the internal actors' requirements because only when their needs are satisfied they can focus on other actors in the meta-organization and not on their own necessities.

3.1.2 Creation of the strategic vision

Being a reflection of the managerial philosophy of the organizational action mode of public services as a whole, this vision represents both a challenge to exceed the *status quo* and the essence of the most significant meta-organizational aspirations in a new *raison d'être*.

The development of the strategic vision focused on Quality offers the opportunity for the management to commit itself to an aggressive strategic approach, but to maintain enough realism to be a permanent source of concentration and motivation.

The organizations of public services are directly influenced by the innovative ability of the services provided, and the approach which reflects the strategic essence of Quality is a major support not only in the articulation of the strategic intention, but also in motivating through means specific to Quality: involvement of the strategic management, continuous improvement, innovation and reengineering at the level of the meta-organization.

3.1.3 Projection of a system of quality measurement

The confirmation of the achievement of the strategic vision of Quality is possible only after the building of a *system of continuous measurement, gathering and reporting of actions referring to the quality of exits and results*. Focus on Quality in the public sector can be situated at the three levels: *the micro, the meso and the macro level of the society.*

Micro-quality is an internal quality concept that applies to the interrelationships of the top, middle and base of an organization.

Meso-quality is an external concept that applies to the relationships between producer and consumer, or supply and demand, or provider and user.

Macro-quality is a generic system concept that applies to the relationships between a public, service and the citizenry, and to the relationship between the state and civil society (C. Politt and G. Bouckaert, 1995, p.14-15).

The construction of the system must be oriented to measure:

1. Meso-quality-, i.e. satisfying users / customers exigencies;
2. Macro-quality-, i.e. gain for society / community;
3. Micro-quality- i.e. interventions in the organization, among which there are:
i. the quality of processes, their efficiency and effectiveness;
ii. the expectations and satisfaction of the members of the organization.

The system of quality measurement was projected in the stage that contains a series of performance indicators, performance through which one can determine on a value scale up to which level public interests have been satisfied and at what expense. The performance of a public organization is determined according to the way human, material, informational and financial resources are used to achieve the objectives established at the level of service beneficiaries' expectations.

The process of measuring the performance results in the public sector is, most specialists admit, an especially difficult one for several reasons:

- *multitude and diversity of stakeholders* in a public institution: current and potential customers, electors, elected representatives, non-profit organizations, professional groups, unions, public managers, the government etc;
- differences in evaluation and perception of performance among stakeholders;
- *non-existence of a competitive environment* where certain services are offered from the position of monopoly some public institutions and administrative authorities have on those services;
- *nature of the public service offered;*
- *complexity of the social-political environment* which generates a series of risks with direct influence on performance achievement; and
- *influence of political values.*

The confirmation of achievement of the strategic vision of Quality is possible only after a *system of continuous measurement, gathering and reporting of actions referring to the quality of exits and results* is built. As you can see in figure 4, the construction of the system must be oriented to measure:

1. short term results, i.e. satisfying users / customers exigencies;
2. long term results, i.e. gain for society / community;
3. the quality of interventions in the organization, among which there are:
i. the quality of processes, their efficiency and effectiveness;
ii. the expectations and satisfaction of the members of the organization.

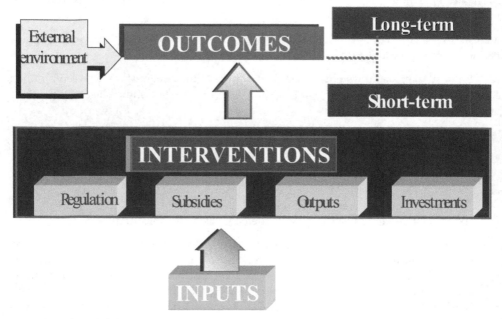

Fig. 4. The chain of the results

The process of measuring the performance results in the public sector is, most specialists admit, an especially difficult one for several *reasons*:

*   *multitude and diversity of stakeholders* in a public organization: current and potential customers, electors, elected representatives, non-profit organiations, professional groups, unions, public managers, the government etc;
*   *differences in evaluation and perception of performance among stakeholders;*
*   *non-existance of a competitive environment* where certain services are offered from the position of monopoly some public institutions and administrative authorities have on those services;
*   *nature of the public service offered;*
*   *complexity of the social-political environment* which generates a series of risks with direct influence on performance achievement;
*   *influence of political values.*

Starting from these ground reasons, in 1994 Stewart and Walsh considered that the evaluation of the performance in the public sector is based on the political argument that defines the coordinates for the identification of performance criteria.

The steps towards performance measuring are the following:

- Defining the expected results (ER);
- Selecting the performance indicators that measure progress in achieving the results;
- Determining final and intermediary performance targes;
- Developing a plan to measure performance using ER indicators;
- Developing tools to gather data referring to each indicator;
- Continuing to gather data and comparing them to the targets previously determined.

Indicators used in defining the expected results (ER) are characterized by:

- Unit of analysis;
- Existing baseline information;
- Useful benchmarks for comparison;
- Expected perceptions or judgments of progress by stakeholders;
- Detailed description of expected conditions or situations to be observed.

Measuring the satisfaction of the metaorganizational actors

From the perspective of external actors, innovation of the public service through quality implies satisfaction of the indicators considered representative for defining the expected results[3]

The fact that external consumers' satisfaction is expressed according to the quality of products and services is the result of a significant number of elements determined by both the product itself (Q) and the circumstances it was delivered in (VALITYM) and thus, we consider that *the measure of the external consumers' satisfaction* (SCE) can be expressed by an algebraic sum of the measure of the satisfaction brought to consumers by each element of the set {QVALITYM} according to relation (1.1):

$$SCE = \sum_{i=1}^{8} Si \quad i \in \{QVALITYM\} \tag{1.1}$$

where Si is the measure of the consumer's satisfaction brought by the quality of element i.

According to this relation, SCE is maximum if each $S_Q \ldots S_M$ is maximum. Practically, problems should be approached differently because the importance of each element of {QVALITYM} varies according to each consumer or segment of consumers.

In this hypothesis, external consumer satisfaction must be expressed according to:

1. Satisfaction (Si) of the quality achieved by each element of the set {QVALITYM};
2. Importance W_i given by the consumer to each element of the set {QVALITYM}.

Thus, relation (1.1) becomes relation (1.2):

---

[3] J.Kelada defines quality through a number of seven indicators : Conformity, Volume, Administrative Procedures, Location, Interrelations, Image, Time Yield (QVALITY). L.G.Popescu adds environment (M) to those defined by J.Kelada.

$$\text{To point out} \quad SCE = \sum_{i=1}^{8} WiSi \quad i \in \{QVALITYM\} \tag{1.2}$$

The importance given by consumers to each W_i, $i \in \{QVALITYM\}$, we score it from 0 to 100 with the restriction:

$$\sum_{i=1}^{8} Wi = 100\%; \quad i \in \{QVALITYM\} \tag{1.3}$$

In figure 5 one can see the importance given by the external actors to each indicator that defines the satisfaction of the public interest.

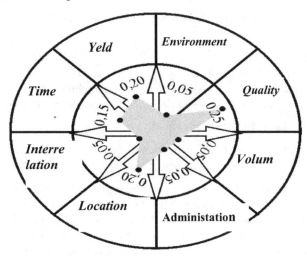

Fig. 5. Illustration of public interest satisfaction according to metaorganizational actors' assessment

The zone in the centre of the figure represents the way the actors of the analysed metaorganization relate to each of the elements of the set QVALITYM, theoretically being considered as having the same importance (the 8 sectors in the circle are identical).

From the analysis of this figure we can conclude that the most important dimension is Quality, it has 25%. Economic efficiency and Delivery venue are the next important dimensions, they have 20%, while delivery Time has 15% and Volume 10%. Consumer-provider relationships and Environment have the same procentage, 5%.

If this exercise is done again on a different type of service, the dimensions in figure 5 will be completely different, being determined by the preferences of the actors of the metaorganization.

Measuring the satisfactionof the internal actors

The relationships between *the internal and external actors* are especially important in the philosophy of Quality Management having a great impact on the satisfaction of the public interest. Although, when we talk about quality there is an instant tendency to refer to the quality of the service, for Quality Management the emphasis on *quality of the people* is substantial. Practice proves that organizations where public managers give importance to the quality of the human relationships have already completed half the way to the achievement of the quality strategic approach. That is why the measurement of their satisfaction is mandatory.

Process measurement

The system of Quality Measurement involves controlling both process results (the so-called final inspection) and control points in the on-going process (the so-called interphasic inspection).

Generally, the achievement of a coherent Measurement System which comprises both types of measurement is difficult. Its difficulty is not to determine a model for the Measurement System, but to mobilize the members of the organization to involve in identifying and measuring in the decisive points, both final and interphasic Starting from these reasons, we can consider that the evaluation of the performance in the public sector is based on the political argument that defines the coordinates for the identification of performance criteria.

3.2 Re-structuring the meta-organization

Participation of the actors contained by the meta-organization to the decision making process regarding restructuring of the organization gives it more legitimacy and increases their satisfaction, as well. The details of restructuring are presented in the following lines.

Participation of the actors contained by the metaorganization to the decision-making process regarding restructuring of the organization gives it more legitimacy and increases their satisfaction, as well. The details of restructuring are presented in the following lines.

3.2.1 Creation of a model of continuous improvement of economic profitableness

In the context of limitation of other economic models, there are questions such as: Which is the real relationship between quality and profitableness? What actions need to be taken to link the process of quality improvement to profitableness? The answers to these questions include decisions referring to: *where and how improvement efforts must be focused, what products / services must be sold, what prices must be charged and when.*

The model we suggest brings up again the paradigm associated to the present economic concepts and, at the same time, it offers a logical solution to build an infrastructure that favours financial performance. This model changes the approch towards increasing the viability of the business and completes the present environment from a global perspective of profit problems.

The development of the model is focused on making the business profitable, i.e. optimizing the two factors: profitableness and assets; which are part of a profitable business.

Profitableness is the profit limit achieved according to the turnover. Profit expresses *the economic value* of the business on the market, respectively its quality to create supplementary wealth. In a broad interpretation of the definition of profitableness, we can say that *it is not measured only by the presence of profit, but alsoby the quality of the business to provide customers with products / services they need and the competition cannot offer.*

The financial health of a business implies placing under discussion the second factor of the relationship, *capital rotation.* The business is more performant when it mobilizes less capital to achieve the same turnover and reduces the time between expenses and cashings. In other words, the faster the answer to the market is, the higher the probability of optimizing the business profit is.

Capital rotation is the speed at which the business converts expenses into turnover, measures the speed at which the business is able to satisfy the market. This rotation depends on the cycle of development, production, and marketing and on the flexibility of the business reaction speed. Thus, the relationship of profitableness although it seems to be a purely financial on, actually is *the measure of the quality of the organizational adaptation to the market.*

In conclusion, the economic model suggested above intends to accurately record *the customer's voice* regarding the added value and the speed at which this is brought onto the market.

The projection of this model implies four stages with the following content:

a.   *analysis of the profitableness of the product / service beginning with the strategic placement;*
b.   *analysis of the present profitableness of the product / service;*
c.   *identification of constraints;* the importance of identifying these constraints and their impact on the strategic objectives lies in determining the directions of improvement efforts. Moreover, the profitableness of the organization would be maximum, at least theoretically if one knows and masters these constraints and limitations. According to their origin, there are several categories of constraints:
i.   behavioural;
ii.   managerial;
iii. production capacities;
iv. logistics;
v.   conditions outside the organization;
vi. mentality regarding costs.
d.   *allocation of resources to different activities (ABC / ABM)*

Creation and reprojection of the allocation of resources system implies focusing on the targets of the business processwithin the chain of value. Each process is divided into component activities to which the corresponding costs are added in the ABC technique (Activities Based on Costs). This technique provides accurate information about the improvement of results and elimination of difficulties. According to this approach resources are allocated to the activities leading to the product achievement, not to the product, as it happens in the traditional approach. Essential to establishing the real costs of projection, execution and distribution of a product, ABC/ABM techniques (Activity Based on Costs/Activity Based on Management) are a useful tool in Quality Management.

Through the association between ABC/ABM techniques and certain practices specific to Quality Management, the process of establishing the total cost of a product goes through the following stages[4]:

a.   *identification of all activities* which, directly or indirrectly, contribute to the achievement of the product is characteristic to technique ABC.
b.   *selection of activities* according to their contribution to the achievement of the product and which expresses technique ABM logics. And (according to Law Pareto) as about

---

[4] Kelada J. op. cit. p. 377

20% of all activities create 80% of the value of the product, they must be paid the greatest attention.

c. *gathering the activities inside the process of achieving the product / service.* Through the practice specific to Quality Management of creation of horizontal teams, they succeed in integrating the activities and the functions into a unitary process.

d. *global integration* by linking the process of achieving the product / service to the internal actors, external ones and stakeholders composing the metaorganization; which reflects the integration of the strategy of Quality at the level of this structure.

3.2.2 Reprojection of work architecture

Public organizations where direct quality costs exceed 15-20% of the annual turnover are considered by O.Gelinier[5] *ghost factories.* Under this name a duality specific to non-performant companies is hiding: on the one hand, *the competitive factory* which produces the economic value, and on the other hand, *the ghost factory* which produces non-value at the expense of the competitive factory.

The implementation of a reduction programme for non-quality based on the principle of *zero-faults* (zero faults, breakdowns, delays, stocks and bureaucracy) is the Quality strategic alternative to eliminate *the ghost factory* and thus, to increase productivuty and production capacity.

The decision of implementation of a reduction programme for non-quality took into account the ghost sub-activities that existed in different sectors of the company and was firstly directed towards the areas with the biggest loss.

Simplification of individual processes through reduction and elimination of *ghost sub-activities* contributes to the reduction of the possibilities for errors to occur and the reduction of errors determines the reduction of the number of repetitive stages in a process, and so on.

According to the previous statements, the methodology of achieving a new work configuration includes the alternative of *continuous improvement* of the existing process. In case this alternative becomes either impossible or achievable or too expensive, the alternative is the implementation of an improved project of the process, through *reengineering*[6]. In this respect, we underline that both alternatives are components of *the improvement process* specific to innovation through quality.

Reconfiguration of the entire network which will integrate the objectives and measurement systems in a *real production process,* is a long term change for which achievement managerial efforts, costs and resources are involved. A very good example is *just in time (JIT),* which is a rather recent managerial approach focused on the elimination of loss at the level of the entire organization based on the reduction of manufacturing times, stocks and waiting times. Achievement of the proposed objectives is possible only if the management promotes the philosophy *just in time* at the level of the entire metaorganization.

---

[5] Gelinier, O. (1990) *,Stratégie de l'entreprise et motivation des hommes,* Ed. Hommes et Techniques, Paris, p.112.
[6] M., Hammer, J.,Champy (1996) *Reeingineering of Enterprise,* ed. Tehnică, Bucureşti, p.25.

3.2.3 The development of the infrastructure for Quality

The organization must learn to deal with the current pretences caused by the speed of the Quality driven transformation, where the informational flux is a vital element. To sustain such a perspective is possible through. *designing the flexible organizational structures to facilitate the communication and to raise the accountability*

The necessity the new, flexible structure is essential. To maintain the traditional structures leads, on short term, to failure in solving both the current and strategic problems. Through this perspective we notice some of the most significant *landmarks of the new organizational form*, which reflects the imperatives of the Quality strategy.

a. *Decentralization of the structures* – the dynamics of cooperation with all the actors of the meta-organization could lead to structures' changes;
b. *Flattening of the structures* – the communication process is almost paralyzed when it follows the hierarchical path; this wins, however, in efficacy when it follows the succession client – provider, already created by the processes.;
c. *The structure in inter-disciplinary teams* – this structural characteristic creates the transversal solution and represents the better solution when compared to: 1) the traditional structures becoming obsolete; 2) the large amount of information presently existing; 3) the failure of hierarchies. Although they comprise employees at all levels, equality exists inside the interdisciplinary teams due to the commonly shared information. Besides, each member of the team is in its centre, and the private objectives are easily integrated, while the motivations grew stronger.

3.3 Re-vitalizing the portfolio of the organization

Compared to the present challenges coming from communities, the traditional organizations in the public sector neglect almost completely *the definition* of future evolution, being mainly concerned with the adoption of defensive positions in order to minimize loss.

Long-term growth is possible through revitalization of the portfolio, a complex strategic approach focused on the way provided services are defined and dealt with.

3.3.1 The focus on public interest

Revitalizing the portfolio in the context of innovation through quality imposes a new managerial vision on the portfolio, the one of *starting point* in projecting future products / services of the organization.

The focus on public interest, vital to the achievement of the exceptional, is the result of *coincidence* between the distinct components or strengths of the organization and the satisfaction of the public interest defined by the requests of all the actors' component of the meta-organization. Obviously, this *coincidence* is not accidental.

The essence of this focusing lies in the examination of the relationships at the level of the meta-organization from the point of view of the public interest.

First of all, this reflects in the identification of the exigency criteria of the actors within the meta-organization (and not in what the management of the organization considers necessary) and, secondly, in the decision regarding the way the offer will satisfy this request efficiently and effectively.

The actors' behavior is influenced by a series of factors which vary from one type of service to another. Knowing and anticipating these variations imposes limitation of the managerial orientations only to those strategic segments where actors' expectations and present and potential abilities of the organization concur.

The segmentation standard taken into account in this transformation was that of *the homogeneity of the expected benefits and the price at which these benefits are obtained*. The factors that express the public interest are bearers of some messages whose significance can be revealed only through direct involvement and attendance of the actors of the meta-organization.

3.3.2 The invention of new public services

The ways of achieving this exigency are presented in the following lines:

a. *Development of the organizational competences*. This process becomes a priority in answering to messages sent by the actors of the meta-organization (representatives of the public interest). Basic competences are represented by those abilities, capacities and technologies that define the specificity of the organization. Their identification, development and combination create new opportunities of satisfying the public interest. In these circumstances, the responsibility of the management id different regarding both the development of the basic competences and constructing and maintaining a motivational environment where creativity is encouraged. The focus on the basic competences of the organization facilitates expansion towards other services than the traditional ones.
b. *Building alliances / partnerships*. Alliances/partnerships are the way to the selective revitalization of the services based on new competences resulted from the combination between the abilities and capacities of the companies that compose the alliance.

The simplest form of alliance is the opportunist alliance. This type of alliance is determined by two factors: technological progress and market globalization. The need imposed by the conditions of the competitiveness, to align to the technological innovation forces organizations to look for alliances with creative partners, offering their position on the market in exchange for access to innovation.

The motivations of the opportunist alliances are very simple: cost reduction, know-how acquisition and risk sharing. Thus, the role of the management of the organization is decisive in building an alliance. In this case, the significant managerial challenge is giving up borders and supporting free knowledge and ability exchange in the interest of the alliance.

Contrary, however attractive the strategic project of the alliance is, it will fail. At the other end from the opportunist alliance, alliances can represent the emergence to a new economic order where businesses are the knots of a network and the managerial ability and efficiency of the network are the clues of the success. The essence of network creation is collaboration based on the complementary principle. The nature of the collaboration within the network is so well defined that the independence and interests of the individual businesses are not affected.

3.3.3 Redefining public services through technology

Using technology to extend or re-define the portfolio of the services provided is the biggest challenge of the revitalization.

Although nowadays the number of the organizations that succeeded in overcoming these challenges is small, in the following decades it will go up because of the technological boom that will influence *the rules of the game*. The technological involvement in the way public services organizations operate starts with the increase of complexity requests for isolated activities and continues with connection and integration of tasks and processes, creation of internal networks and global redefinition of the public services offer.

Technological advantages imply: integration of processes through technology, involvement of technology in process reengineering, involvement of technology in network design. The informational technology next to consumers' needs and organizational abilities form the triad of events that when they are combined create opportunities for revitalization.

Technology *links* the actors of the meta-organization and its different departments facilitating common progress and improving speed and efficiency in finality achievement.

One of the most important technological applications is its strategic contribution to gathering and developing knowledge and to the acceleration of their dissemination within the meta-organization.

3.4 Re-inventing the mentalities of the actors

At the center of the Quality cycle lays the human aspect (see figure 4). No activity can be carried out effectively if the people involved are not willing to cooperate.

In order to do so, they have to be convinced that what they are asked to do is for their own benefit, rather than for another person or group of persons, be it the stakeholders or the customers.

3.4.1 The new quality culture

In order to achieve the strategic quality approach, an organization has to change. This change goes far beyond altering a method or modification of a process. It is, at first, a change in culture that is required; all other major changes will then follow from that.

Generally, culture represents the way through which the members of a group communicate, both among themselves and with other groups within the company. In the context of the communication process there are highlighted common behaviors, customs, practices, values and beliefs.

The new quality culture is characterized by the expanding of empowerment within the organization, on the basis of the market-in concept, by which the customer concept is introduced inside the public organization; thus, the public servants become binders of the institution, each of them knowing who he works for and knowing that person's exigencies, things which contribute to the creation and preservation of the customer cult in the entire public organization.

The culture is a vision of the future and a set of mostly unwritten values. It is partly inherited, partly effected by all actors who exercise an influence on and in the organization (J. Kelada, p.187).

Ideally, actors of the meta-organization all share the same long-term global vision of the organization and of the environment in which they have to live and thrive. Consequently, every action they take will be inspired by and geared to this vision.

Culture is not a static concept or reality; it evolves with time and changes, often gradually but sometimes noticeably and abruptly. Cultural changes result from a constant feedback whether formal or informal conscious or unconscious - from the results of the culture (J. Kelada, p.189).

The importance of the new quality culture, articulated around transforming the mentality, consists, especially, of the implications due to it upon putting into practice the changing project. In other words, the success of changing the organizational culture depends on the total involvement of the public management, carefully focused on the institution's clients, but also on the public servants.

3.4.2 A new style of management – The Total Quality Management

TQM is principally based on a managerial philosophy, a way of thinking from which emanates a way of doing that is assumed by all persons in the organization as well by its external partners. This philosophy is customer focused, people centered, partner assisted, and environmentally conscious.

The Total Quality Management is an approach through which the organization makes sure that its processes, products, and services contribute to the achievement of clearly defined results. This managerial approach is mainly characterized by:

- Encouragement of stakeholders' participation;
- Realistic definition of expected changes or results;
- Selection of performance indicators;
- Risk assessment;
- Gathering information about performance;
- Using this information in the decision-making process;
- Relation to performance.

Based on the quality strategic approach, TQM has three dimensions: (1) the human dimension: political and psychological; (2) a logical dimension: rational and systematic: and (3) a technological dimension: mechanical and systematic. There are opinions through scholars that the first is the most important (J. Kelada,, p.57).

The human aspect has two objectives: to start and to maintain a quality process in the meta-organization. Indeed, top managers must be convinced that they have to change their ways of doing public business; they have to exercise strong leadership inside the organization with their people and outside with the external actor of meta-organization, both of whom they have to mobilize.

They have to "walk their talk", practice what they preach, be committed and involved. Then, once the strategic quality approach is under way, they have to continuously reinforce it by their attitude and behavior, by rewards and recognition, by participation and teamwork. The new public managers must not only acquire innovative and creative abilities necessary to define new horizons rapidly, but also facilitate their putting into practice.

In this context, the contribution to behavioral re-orientation of organizational actors towards environmental changes is one of the vectors of the new managerial orientations.

The challenges of the current institutional climate are, in my opinion, generated not so much by the lack of knowledge of the rules of the game, but mostly by their continuous changing. The new managers in the public sector must not only rapidly accumulate innovative and creative capabilities necessary for defining certain new horizons, but also to facilitate their transposing into practice.

In this context, the contribution to the behavioral reorientation of the public servants in accordance with the environmental changes is one of the strategic vectors of the current managerial approach. In terms of consequences, it is necessary to give up the old paradigms and the total commitment, in spirit and in action, in a changing process defined on the long term, in the sense of the new public management exigencies.

3.4.3 Building a continuous learning system

The learning system represents the most advanced way of renewing the mentalities of the public organization's members. The promotion of this system implies the undertaking by the public organizations' managers of a double responsibility: (1) acknowledging the employees' individual needs for personal and professional development and (2) creating opportunities for satisfying these needs. The professionalism of public institutions has a rising dynamics, directly proportional to the number of public servants who continuously redefine their roles and responsibilities on the basis of the permanent improvement and adaptation of the public activities.

However, there are numerous examples of public organizations that ignore, willingly or out of a misunderstanding, this unlimited potential offered by each employee. Even more, the current circumstances indicate that, while bureaucracy is expanding its dimensions and complexity, the construction of a learning system becomes impossible. The reason at the foundation of this statement is argued further.

Firstly, those who represent the existing paradigms do not have the availability necessary for understanding the need for change and, implicitly, they are refractory to anything that takes them further from the "old road".

Secondly, every bureaucrat's mission is defined in terms of self-sufficiency and self-commitment; this signifies the absence of the collective contacts, lateral or that cross the organization. The adopting of certain decisions and executive dispositions in this sense would not have the benefic effects taken into account because the majority of the factors involved in the decisional process are the representatives of the old paradigms.

4. Conclusion

Accordingly, we think that the metaorganization built corresponds to the exigence imposeed by sector IV of figure 6. It integrates co-production and co-evaluation, but adds also co-design and co-decision. These two crucial participative steps are only possible if there is a combined external and open orientation with a focus on outputs and outcomes. This results in co-governing.[7]

---

[7] C., Pollitt, G.,.Bouckaert, E.,Loffler, *Making Quality Sustainable: Co-Design, Co-Decide, Co-Produce, and Co-Evaluate*, The 4QC Conference, Tampere, 2006., p. 18.

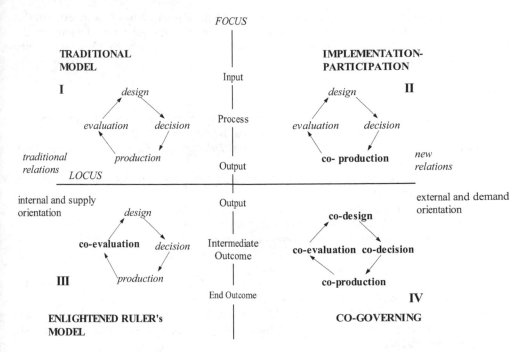

Fig. 6. Models for sustainability in the public sector Sursa: C., Pollitt and G.,Bouckaert, Quality Improvement in European Public Services, 1995, p.163

There are three important aspects in the governance of complex public sector networks. The first aspect is the *context*- defined as the environment. Second is *the complexity* – defined as the number and variety of the system's elements and the relations between the system elements. The third aspect is *governance*- defined as directed influencing.

In a network of many separate actors, with different and often conflicting goals and interests and with diverging power positions, no single dominant actor exists. Such complexity means negotiation and implies a different form of governance than mono-centric, mono-rational, hierarchical top-down control by an omnipotent government. On the other hand, public governance in complex network differs from the polar opposite of hierarchy, total autonomy of actors. Networks are characterized by many dependencies and relationships among the actors.

The distinction between a multi-actor network and completely autonomous actors is not without meaning. It means that actors not entirely independent and it also means that although actors are not hierarchically sub-or super-ordinate, they are not completely equivalent. Government will always take a different position than other societal public and private actors in a network. Government cannot dominate and unilaterally dictate but nevertheless, it is not entirely equivalent to all other actors. This is not a normative statement but an empirical observation than the role of government in networks is special and unlike the roles of many other actors. This does not imply a return to top-down control. It does imply that full horizontality and total autonomy of actors is an unrealistic model of a public policy network.

Network governance concept lies somewhere in the grey area between both extremes of hierarchy and market. It is remarkable that the few people interest ed in developing the public sector-oriented approach all emphasize the importance of public *governance* for public management and public organization.

5. References

Ansoff, I., E. McDonneell, 1990, *Implanting Strategic Management*, Prentice Hall, New York ;

Drucker, P, 2000, . Post capitalist Society, ed. Image, Bucharest

Fukuyama, F., 2004, *State –Building. Governance and World Order in the 21st Century*, Ed. Antet XX Press, Bucharest

Gelinier, O. 1990, *Stratégie de l'entreprise et motivation des hommes*, Ed. Hommes et Techniques, Paris

Gilles, P., 1994, *New Patterns of Governance*, Kenneth Press, Ottawa;

Guskin, A., 1994, *Part II: Restructuring the Role of the Faculty*, Change 26, no.5: 16-25.;

Hammer, M. J.,Champy ,1996, *Reeingineering of Enterprise*, ed. Tehnică, Bucureşti

Harvey L., 2000, *New Realities: The Relationship between Higher Education and Employment*, Source: Tertiary Education and Management ;

Harvey, L., Knight, P.T., 1996, *Transforming Higher Education.*, Buckingham: Society for Research into Higher Education (SRHE) and Open University Press;

Harvey, L., Moon, S.,Geall, VGraduates, 1997, *Organizational Change and Students' Attributes.*, Birmingham: Centre for Research into Quality (CRQ) and Association of Graduate Recruiters (AGR)

J Taylor 2003, *Institutional Diversity in UK Higher Education: Policy and Outcomes Since the End of the Binary Divide* - Higher Education Quarterly, - Wiley Online Library

Lomas, L. , 2007, *Zen, motorcycle maintenance and quality in higher education* , Quality Assurance in Education, Volume 15 No 4, pp402-412

Morgan, G, 1989, *Creative Organization Theory*, Newbury park, California, Sage Publication;

Pirsig, Robert M. , 1974, *Zen and the Art of Motorcycle Maintenance – An Inquiry into Values*, Boadly Head, UK.

Politt C., and G., Bouckaert, 1995, *Quality Improvement in European Public Services*, SAGE Publication Ltd. London

Pollitt, C., G.,.Bouckaert, E.,Loffler, 2006, *Making Quality Sustainable: Co-Design, Co-Decide, Co-Produce, and Co-Evaluate*, Report of The 4QC Conference, Tampere

Popescu, L.G. 2007, *The innovation of the public services by quality strategic approach*, Transylvanian Review of Administrative Sciences nr.19(E).

Popescu, L.G., 2005, *The company managed by Total Quality*, International Conference RaDMI , Belgrad, pp. 230-238.

Popescu, L.G., 2011, Institutionalization of Strategic Responsiveness in Higher Education, Report of The ISQM Conference, Sibiu

Stein, F.R. 1999, *The next phase of Total Quality Management*, Marcel Dekker Inc., New York

Teece J.D., 2009, *Dynamic, Capabilities and Strategic Management*, Oxford University Press

Woodhouse, David, *Introduction to Quality Assurance*, 2009, AUQA

Section 2

Quality Management Practices

4

Service Quality Dimensions in the Banking Industry and Its Effect on Customer Satisfaction (Case Study)

Soleyman Iranzadeh[1,*] and Farzam Chakherlouy[2]

[1]*Department of Management, Tabriz Branch, Islamic Azad University, Tabriz,*
[2]*Tabriz Business Training Center, East Azerbaijan, Tabriz,*
Iran

1. Introduction

In the competitive world, it is a key factor to meet customer satisfaction which is followed by organizations. Customer satisfaction is one of fundamental principles of quality management. Loyal and satisfied customers bring about stable income for organizations. Thus organizations pay special attention to factors such as customer knowledge, customer relationships, and determination of methods for meeting customer satisfaction and for providing suitable goods and services to meet customer needs because Customer is the most important asset of organization. Peter Drucker claims that customer satisfaction is the goal and aim of all activities. "Increased competition, meeting customer satisfactions …are new concepts that have strongly affected current world in a way that one cannot compete or even survive according old ideas in new world." There is no doubt that development of technology has increased customers 'expectations to receive quality and on time services. Customers will no longer accept any quality of services. Service quality is increasingly becoming a major strategic variable (Robledo, 2001; Terziovski and Dean, 1998). And This construct has received increased scrutiny during the last few decades (Svensson, 2004). In the 1980, large organizations became more interested in the development of service quality measures (Dedeke, 2003). Much of the research has focused on measuring service quality using the SERVQUAL instrument (Kang, 2006; Ladhair, 2008). While, the SERVQUAL technique has attracted a lot of attention for its conceptualization of quality measurement issues, it has also attracted criticism (O'Neill *et al.*, 1998). One criticism of SERVQUAL has been the point that the instrument mainly focuses on the service delivery process. That is, while the contemporary studies on service quality seemingly focused on the process of service delivery, additional aspects to be considered have already been suggested, especially by European scholars (Kang and James, 2004). For example, Gronroos noted that the quality of a service as perceived by customers has three dimensions: functional (or process) dimension, technical (or outcome) dimension and image (Kang and James, 2004). Thus, the European perspective versus

---

* Corresponding Author

American perspective suggests that service quality considers two more components, technical quality and image. No efforts have been made to test the European perspective to determine the dimensions of services quality in banking industry of Iran. The purpose of this study is to extend our understanding of service quality by empirically examining the conceptualization of service quality suggested in the European perspective (i.e., Gronroos's model) (Kang and James, 2004). Researchers have tried to develop conceptual models to explain the service quality and to measure consumers perceived service quality in different industries (Seth *et al.*, 2005). A good operational example of a standardized framework for understanding service quality is the SERVQUAL instrument developed by Parasuraman. The researchers discovered five general dimensions with focus group interviews which they labeled: reliability, responsiveness, tangibles, assurance and empathy (Wong and Sohal, 2002). Service quality is an important factor for success in the banking sector. Thus, some bank managers emphasize the various dimensions of service quality (Glaveli *et al.*, 2006).

Parasuraman and his supporters introduce 5 general dimensions of service quality of any service provider. These dimensions include:

1. Tangibility: physical facilities, tools, machines, personnel, materials and communication channels.
2. Trustworthiness: the ability to provide promised services in a proper and reliable way.
3. Accountability: to have the interest in providing appropriate service and generally helping customers.
4. Reliability: knowledgeable and polite personnel and their ability to win customers trust and confidence.
5. Sympathy: taking care and paying attention to individuals

Thus any bank that fails to surpass customer expectations and meet customer satisfaction will not be able to compete with other banks. It is the most difficult challenge for a bank that its customers transfer their accounts to rival banks because of better services.

Currently, Iranian banking industry has a dynamic and complex environment. Increasing development of public and private commercial banks and improvements in the kind and way services have been offered, increasing expansion of financial and credential institutions and organs to provide financial and non financial services, and increasing development of technology in banking industry offers a competitive and special environment to any organization. It requires that active organization in the banking industry of country pay more attention to customer satisfaction. This goal will not be achieved without localized models and indicators through which one can make sure of customer satisfaction. However, appropriate studies have not been done to realize dimensions of service quality. So this study aims to estimate and evaluate indicators of Parasuraman model of service quality through a survey from customers of Melli bank to have a better understanding of these dimensions.

We intend to evaluate five dimensions of service quality according Parasuraman model.

So we will try to answer following questions about Bank Melli Iran of East Azerbaijan:

1. How does sympathy influence customer' satisfaction in Bank Melli Iran of East Azerbaijan?
2. How does reliability influence customer satisfaction in Bank Melli Iran of East Azerbaijan?
3. How does accountability influence on customer satisfaction in Bank Melli Iran of East Azerbaijan?
4. How does tangibility of services influence on customer satisfaction in Bank Melli Iran of East Azerbaijan?
5. How does validity of provided services influence on customer satisfaction in Bank Melli Iran of East Azerbaijan?

2. Research objectives

This research aims at one main objective and five subsidiary objectives that are discussed as following:

Main objective

To study dimensions of service quality and to evaluate the effects of these dimensions on satisfaction of Melli bank customers in the province of Azerbaijan-e Sharghi.

Subsidiary objectives

- To study the effect of sympathy against services provided on the satisfaction of Melli Bank customers in the province of Azerbaijan-e Sharghi.
- To study the effect of reliability of services provided on the satisfaction of Melli Bank customers in the province of Azerbaijan-e Sharghi.
- To study the effect of tangibility of services offered on the satisfaction of Melli Bank customers in the province of Azerbaijan-e Sharghi.
- To study the effect of trustworthiness of services offered on the satisfaction of Melli Bank customers in the province of Azerbaijan-e Sharghi.
- To study the effect of being responsible for the services offered on the satisfaction of Melli Bank customers in the province of Azerbaijan-e Sharghi.

Research hypotheses

1. Sympathy has above average effect on customer satisfaction in Bank Melli Iran of East Azerbaijan.
2. Reliability has above average effect on customer satisfaction in Bank Melli Iran of East Azerbaijan.
3. Tangibility has above average effect on customer satisfaction in Bank Melli Iran of East Azerbaijan.
4. Trustworthiness has above average effect on customer satisfaction in Bank Melli Iran of East Azerbaijan.
5. Accountability has above average effect on customer satisfaction in Bank Melli Iran of East Azerbaijan.

Variables

Independent variables: Sympathy, tangibility, trustworthiness, reliability and accountability of provided services Dependant variables: Bank customer satisfaction

Analytical model

Research scope

1. **Subjective Scope:** to evaluate the effect of dimensions of quality of services in Banking industry on customer satisfaction according Parasuraman model in Bank Melli Iran of East Azerbaijan.
2. **Spatial Scope:** Iran, Bank Melli Iran of East Azerbaijan
3. **Chronological Scope:** around 8 months that begins from July 2010 and ends at December 2010.

Statistical society

Statistical society of this study is all customers of Bank Melli Iran (East Azerbaijan) who were present in the bank (in East Azerbaijan).

Proposal

The research design used in this study is Descriptive research design. In terms of strategic purpose, this study is causative and survey research is applied to collect data. Thus, some books and documents have been studied to collect necessary information. This study also benefits from library studies, questionnaires, notes- to prepare theoretical framework. Therefore dimensions of quality of services in banking industry and its effect on customer satisfaction have been detected as well as demographic situation of all individuals in statistical sample. Assessments tools have been utilized for structural quality of services and then customer satisfaction has been evaluated. This study uses field work to collect experimental data. To extract and analyze experimental data- according to descriptive research design- various statistical methods have been applied to study existing situation of sympathy, accountability, reliability, tangibility and trustworthiness. Advanced statistical analysis such as T-student test has been used to detect and evaluate the effect of variables.

3. Society, sample and sampling method

The society of this study comprises of all customers of Bank Melli Iran and its branches in East Azerbaijan. Simple random sampling has been applied so all branches of society have the same chance of selection. Samples have been selected in random and the questionnaires are distributed among the customers randomly. The questionnaire has been distributed among 300 customers from whom 256 customers completed and gave back the questionnaire. It shows that distribution of questionnaire among customers of Bank Melli Iran (East Azerbaijan)- as statistical society of this study- has been randomly. The number of statistical sample according sample size formula determined to be 267.

$$n = \frac{z_{\frac{\alpha}{2}}^2 * p(1-p)}{e^2} = \frac{(1.96)^2 * 0.5(1 - 0.5)}{(0.06)^2} = 266.77 \approx 267$$

4. Validity and reliability of measurement tools

As the validity and reliability of the questionnaire has already been evaluated in another study by the author, we will suggest existing test results for a questionnaire of 43 questions and then we propose reliability test for current study by a questionnaire of 35 questions.

Standardized questionnaire has been used to collect data. It is the same questionnaire which has been used by King& James in 2004 in Europe. The questionnaire has been translated into Persian then its validity and reliability has been tested and finally it has been used. It includes 35 questions which have been designed according to Likert Spectrum. Questions are divided into 5 parts. Questions of every part are used to evaluate one of features (variables). Structural way has been used to determine validity of tools. Cronbach's α (alpha) coefficient calculation is used to determine the reliability of scales. The results indicate the reliability of measured features. A summary of results for validity and reliability of tests is indicated in table 1.

Scales	Cronbach's α (alpha) coefficient	Sampling adequacy	Bartlett sphericity test
Trustworthiness	0/78	0/75	206/113
Accountability	0/80	0/69	221/696
Reliability	0/84	0/69	200/996
Sympathy	0/85	0/60	33/860
Tangibility	0/79	0/78	318/818
Customer satisfaction	0/78	0/75	206/113

Table 1. Summary of Results for Validity and Reliability Test

With regard to results of validity and reliability tests of previous study, validity of current questionnaire has been evaluated by formal method. First of all, the questionnaire has been distributed among 35 customers of statistical society then the answers for every question have been assessed separately and the percentage has been calculated for every question. Firstly, the percentage for every question is above 90% of sample size. Secondly, all questions have been responded by maximum number of customers in an expected way. In addition to theoretical basis, corrective comments of respondents and expertise were considered to confirm the content validity of questionnaire in primary phase. It is clear that after pilot questionnaire, the answers have been evaluated by Cronbach's α coefficient (SPSS) - a reliable tool to determine the reliability of questionnaire. Using the results of

Cronbach's α coefficient (SPSS) the reliability of questionnaire was determined to be over 83% (it means 0.836). Thus the validity and reliability of questionnaire is confirmed.

Cronbach's Alpha	Cronbach's Alpha Based on Standardized Items	N of Items
.836	.836	35

Case Processing Summary

		N	%
Cases	Valid	35	100.0
	Excludedª	0	.0
	Total	35	100.0

a. Listwise deletion based on all variables in the procedure.

Table 2. Calculation of Reliability of Questionnaire

5. Assessment tools and data collection method

A questionnaire has been designed according to existing literature to evaluate the dimensions of quality of services and its effect on Bank Melli Iran (East Azerbaijan) customer satisfaction. The indicators have been assessed through consideration of research background to evaluate five dimensions (It is reflected in analytical model of study). It is noteworthy that variables have been assessed by 5 point Likert scale. A questionnaire of 35 questions asks Bank Melli Iran (East Azerbaijan) customers to answer every question according a rating scale (Likert spectrum) starting from very low, below average, average, above average, very high. Assessments tools are 5 hypothesis of study.

5.1 Data analysis method

This study applies T-student test to test the hypothesis. The effect of any hypothesis will be evaluated separately. Significance level is 0.05 (degree of confidence: 95%) to prove or disprove the hypothesis. The questionnaire includes 35 questions. Each group of questions of hypothesis (1-5) includes 6 question of questionnaire according table 3.

questions	Summarized questions	Questions of questionnaire
1	the effect of sympathy on customer satisfaction	1-7
2	the effect of trustworthiness on customer satisfaction	8-14
3	the effect of tangibility on customer satisfaction	15-21
4	the effect of reliability on customer satisfaction	22-28
5	the effect of accountability on customer satisfaction	29-35

Table 3. Questions of Questionnaire

Table 4 presents the relative frequency of respondents. The results could be presented in a descriptive way for any question of study.

Levels	Percentage				
	Question1	Question 2	Question 3	Question 4	Question 5
Very low	12.67	10.72	10.83	12.28	12.17
Below average	16.43	20.31	21.04	18.03	16.63
Average	29.58	28.01	29.07	28.74	30.14
Above average	24.20	24	22.43	24.89	24.83
Very high	17.10	16.96	16.63	16.06	16.23
Total	100	100	100	100	100

Table 4. Relative Frequency Distribution of Respondents

Analysis of first hypothesis

One-Sample Statistics

	N	Mean	Std. Deviation	Std. Error Mean
sympathy	256	3.16	.492	.031

One-Sample Test

	Test Value = 3					
					95% Confidence Interval of the Difference	
	t	df	Sig. (2-tailed)	Mean Difference	Lower	Upper
sympathy	5.101	255	.000	.157	.10	.22

Table 5. First hypothesis testing, results of data analysis (t-student test): sympathy against provided services

H_0: The effect of sympathy against provided services on customer satisfaction in Bank Melli Iran (East Azerbaijan) is not higher than average.

H_1: The effect of sympathy against provided services on customer satisfaction in Bank Melli Iran (East Azerbaijan) is higher than average.

$$\begin{cases} H_0 : \mu < 3 \\ H_1 : \mu \geq 3 \end{cases}$$

$$\begin{cases} \alpha = 0/05 \\ T_{(0/05,255)=1.66} < T_{cal} = 5.101 \end{cases}$$

$T_{cal} (= 5.101) > T_{(0.05,255)}$ (= 1.66), sympathy index average is 3.16 -according to respondents- which is higher than supposed index (Test Value= 3) significance level α = .000 (which is lower than minimum significance level). The difference between two indices is remarkable. So it cannot be by accident and we cannot omit it. Thus we can disprove H_0 with significance level of 0.05 and prove H_1 with a degree of confidence= 0.95. This degree of confidence shows that 12.67 percent of Bank Melli Iran (East Azerbaijan) customers find the effect of sympathy against provided services to be very low, 16.43 percent of customers find it below average, 29.58 percent of customers find it as average, 24.20 percent of customers find it to be above average and 17.10 percent of customers find the effect of sympathy very high.

Analysis of second hypothesis

One-Sample Statistics

	N	Mean	Std. Deviation	Std. Error Mean
trustworthiness	256	3.16	.520	.032

One-Sample Test

	Test Value = 3					
	t	df	Sig. (2-tailed)	Mean Difference	95% Confidence Interval of the Difference	
					Lower	Upper
trustworthiness	4.830	255	.000	.157	.09	.22

Table 6. Second hypothesis testing, results of data analysis (t-student test): trustworthiness of provided services

H_0: The effect of trustworthiness of provided services on customer satisfaction in Bank Melli Iran (East Azerbaijan) is not higher than average.

H_1: The effect of trustworthiness of provided services on customer satisfaction in Bank Melli Iran (East Azerbaijan) is higher than average.

$$\begin{cases} H_0 : \mu < 3 \\ H_1 : \mu \geq 3 \end{cases}$$

$$\begin{cases} \alpha = 0/05 \\ T_{(0/05,255)=1.66} < T_{cal} = 4.830 \end{cases}$$

$T_{cal}(= 4.830) > T_{(0.05,255)}$ $(= 1.66$), trustworthiness index average is 3.16 -according to respondents- which is higher than supposed index (Test Value= 3) and significance level α = .000 (which is lower than minimum significance level). The difference between two indexes is remarkable. So it cannot be by accident and we cannot omit it. Thus we can disprove H_0 with significance level of 0.05 and prove H_1 with a degree of confidence= 0.95. This degree of confidence shows that 10.72 percent of Bank Melli Iran (East Azerbaijan) customers find the effect of trustworthiness of provided services to be very low, 20.31 percent of customers find it below average, 28.01 percent of customers find it as average, 24 percent of customers find it to be above average and 16.96 percent of customers find the effect of trustworthiness very high.

Analysis of third hypothesis

One-Sample Statistics

	N	Mean	Std. Deviation	Std. Error Mean
tangibility	256	3.13	.521	.033

One-Sample Test

	Test Value = 3					
	t	df	Sig. (2-tailed)	Mean Difference	95% Confidence Interval of the Difference	
					Lower	Upper
tangibility	3.930	255	.000	.128	.06	.19

Table 7. Third hypothesis testing, results of data analysis (t-student test): tangibility of provided services

H_0: The effect of tangibility of provided services on customer satisfaction in Bank Melli Iran (East Azerbaijan) is not higher than average.

H_1: The effect of tangibility of provided services on customer satisfaction in Bank Melli Iran (East Azerbaijan) is higher than average.

$$\begin{cases} H_0 : \mu < 3 \\ H_1 : \mu \geq 3 \end{cases}$$

$$\begin{cases} \alpha = 0/05 \\ T_{(0/05,255)=1.66} < T_{cal} = 3.930 \end{cases}$$

$T_{cal} (= 4.830) > T_{(0.05,255)}$ (= 1.66), tangibility index average is 3.13 -according to respondents- which is higher than supposed index (Test Value= 3) and significance level α = .000 (which is lower than minimum significance level). The difference between two indexes is remarkable. So it cannot be by accident and we cannot omit it. Therefore, we can disprove H_0 with significance level of 0.05 and prove H_1 with a degree of confidence= 0.95. This degree of confidence shows that 10.83 percent of Bank Melli Iran (East Azerbaijan) customers find the effect of tangibility of provided services to be very low, 21.04 percent of customers find it below average, 29.07 percent of customers find it as average, 22.43 percent of customers find it to be above average and 16.63 percent of customers find the effect of trustworthiness very high.

Analysis of forth hypothesis

One-Sample Statistics

	N	Mean	Std. Deviation	Std. Error Mean
reliability	256	3.14	.494	.031

One-Sample Test

	Test Value = 3					
	t	df	Sig. (2-tailed)	Mean Difference	95% Confidence Interval of the Difference	
					Lower	Upper
reliability	4.683	255	.000	.145	.08	.21

Table 8. Forth hypothesis testing, results of data analysis (t-student test): reliability of provided services

H_0: The effect of reliability of provided services on customer satisfaction in Bank Melli Iran (East Azerbaijan) is not higher than average.

H_1: The effect of reliability of provided services on customer satisfaction in Bank Melli Iran (East Azerbaijan) is higher than average.

$$\begin{cases} H_0 : \mu < 3 \\ H_1 : \mu \geq 3 \end{cases}$$

$$\begin{cases} \alpha = 0/05 \\ T_{(0/05,255)=1.66} < T_{cal} = 4.683 \end{cases}$$

$T_{cal}(= 4.683) > T_{(0.05,255)}$ (= 1.66), reliability index average is 3.14 -according to respondents- which is higher than supposed index (Test Value= 3) and significance level α = .000 (which is lower than minimum significance level). The difference between two indexes is remarkable. So it cannot be by accident and we cannot omit it. Thus we can disprove H_0 with significance level of 0.05 and prove H_1 with a degree of confidence= 0.95. This degree of confidence shows that 12.28 percent of Bank Melli Iran (East Azerbaijan) customers find the effect of reliability of provided services to be very low, 18.03 percent of customers find it below average, 28.74 percent of customers find it as average, 24.89 percent of customers find it to be above average and 16.06 percent of customers find the effect of reliability very high.

Analysis of fifth hypothesis

One-Sample Statistics

	N	Mean	Std. Deviation	Std. Error Mean
accountability	256	3.16	.480	.030

One-Sample Test

	Test Value = 3					
					95% Confidence Interval of the Difference	
	t	df	Sig. (2-tailed)	Mean Difference	Lower	Upper
accountability	5.355	255	.000	.161	.10	.22

Table 9. Fifth hypothesis testing, results of data analysis (t-student test): accountability of provided services

H_0: The effect of accountability of provided services on customer satisfaction in Bank Melli Iran (East Azerbaijan) is not higher than average.

H_1: The effect of accountability of provided services on customer satisfaction in Bank Melli Iran (East Azerbaijan) is higher than average.

$$\begin{cases} H_0 : \mu < 3 \\ H_1 : \mu \geq 3 \end{cases}$$

$$\begin{cases} \alpha = 0/05 \\ T_{(0/05,255)=1.66} < T_{cal} = 5.355 \end{cases}$$

$T_{cal}(= 5.355) > T_{(0.05,255)}$ (= 1.66), accountability index average is 3.16 -according to respondents- which is higher than supposed index (Test Value= 3) and significance level α = .000 (which is lower than minimum significance level). The difference between two indexes is remarkable. So it cannot be by accident and we cannot omit it. So we can disprove H_0 with significance level of 0.05 and prove H_1 with a degree of confidence= 0.95. This degree of confidence shows that 12.17 percent of Bank Melli Iran (East Azerbaijan) customers find the effect of reliability of provided services to be very low, 16.63 percent of customers find it below average, 30.14 percent of customers find it as average, 24.83 percent of customers find it to be above average and 16.23 percent of customers find the effect of accountability very high.

Column	hypothesis	Test type	Significance level	Hypothesis condition
1	First hypothesis	T-student	0.05	proved
2	Second hypothesis	T-student	0.05	proved
3	Third hypothesis	T-student	0.05	Proved
4	Fourth hypothesis	T-student	0.05	proved
5	Fifth hypothesis	T-student	0.05	proved

Table 10. Hypothesis Condition

6. Discussion and conclusion

Nowadays, because of the global economic conditions and internal situation in Iran, the necessity of transformation in banking system has become more tangible and definite than any other time. As a competent and efficient banking system plays an effective role in realization of development programs, it is necessary to pay attention to this issue as one of most fundamental programs for development of the country. Iranian banking industry -as one of effective basis in the economy of the country- plays a decisive role in economical activities. So, effectiveness and efficiency of banking activities will play a decisive role in economical development of the country. Having all hypothesis of the stud in mind and accepting these hypotheses, it is important to note that it is rather a continuous program- not only a special program- to improve the quality of bank services. This requires constant attention of banks' senior executives. The existing gap in the productivity of world's banking industry indicates the deep gap between productivity and international standards. This gap arises from several factors. The excess of demand over supply can act as a positive potential to increase the productivity but our failure to employ various techniques in making effective use of banking resources has been resulted in inefficiency of the banking activities. Because allocation of limited resources would not be in a right way, and not only a reasonable profit could not be achieved but also the customers would not be satisfied with bank services. As foreign and private banks has a short history in Iranian banking industry, it is necessary for public banks to pay more attention to development and provision of commercial services and dimensions of service quality that enables these banks to compete continuously with private banks. For this purpose it is necessary for banks to aim at increasing the quality of services with a strategic vision to increase the satisfaction of customers. This will not be achieved unless these banks can realize the effective factors on understanding the service quality and explaining appropriate strategies for each of these factors.

7. References

Dedeke, A., 2003. Service quality: A fulfillment–oriented and interactions–centered approach. Managing Service Quality, 13: 276-289.

Glaveli, N., E. Petridou, C. Liassides and C. Spathis, 2006. Bank service quality: Evidence from five balkan countries. Managing Service Quality, 16: 380-394.

Kang, G.D. and J. James, 2004. Service quality dimensions: An examination of gronrooss service quality model. Managing Service Quality, 14: 266-277.

Kang, G.D., 2006. The hierarchical structure service quality: Integration of technical and functional quality. Managing Service Quality, 16: 37-56.

Ladhair, R., 2008. Alternative measures of service quality: A review. Managing Service Quality, 18: 65-86.

O'Neill, M.A., A.J. Palmer and R. Beggs, 1998. The effects of survey timing on perceptions of service quality. Managing Service Quality, 8: 126-132.

Robledo, M.A., 2001. Measuring and managing service quality: Integrating customer expectations. Managing Service Quality, 11: 22-31.

Seth, N., S.G. Deshmukh and P. Vrat, 2005. Service quality models: A review. Int. J. Quality Relaibility Manage., 22: 913-949.

Svensson, G., 2004. Interactive service quality in service encounters: Empirical illustration and models. Managing Service Quality, 14: 278-287.

Terziovski, M. and A. Dean, 1998. Best perdictors of quality performance in Australian service organization. Managing Service Quality, 8: 359-366.

Wong, A. and A. Sohal, 2002. Customers perspectives on service quality and relationship quality in retail encounters. Managing Service Quality, 12: 424-433.

5

Quality Assurance for the POCT Systems

Luděk Šprongl
Central Laboratory, Šumperk Hospital
Czech Republic

1. Introduction

Point-of-care testing (POCT) is in the fact laboratory testing provided near the site of patient care or directly in this place. POCT includes in the present time bedside testing, near-patient testing, physicians´ office testing, extra-laboratory testing, decentralized testing, off-site, ancillary, and alternative testing, and, of course, point-of-care testing. Laboratorian and clinical staff often use these terms synonymously, but POCT can be delivered in many different ways. Different tests and technique can be also hidden in different terms.

The history of the POCT is relatively old. We can see now very sophisticated techniques and methods. However, in the beginning of laboratory tests was everything in the hands of the practitioners. The number of tests was arising during the time and laboratory tests were moved to the laboratories. Laboratories were higher and higher and time for the results was longer and longer. From these reasons some tests move back near to the patient. The typical and one of the oldest POCT test is chemical examination of the urine on the strip. But more and more tests are common outside of the hospitals and laboratory. Above all we can mention glucose, APTT, CRP, HbA1c, hCG, lactate, drugs or lipid´s profile. And the development of POCT brings new methods – bilirubin, microbiology tests, PTH etc.

The development of the POCT tests depends on many different factors (1). Here are the availability of the tests, the price, patient´s requirements for the tests and the technology development. However it´s clear, the main factor is the last factor (technology development). This development enables enlargement test´s spectrum for the POCT. New technologies make possible to provide a lot of tests not only in practitioner´s office or in the terrain, but also at home with the connection via internet to the doctor. The wide of the new tests in the hospital POCT is limited, as the majority of laboratory tests is in the right time with the better price and quality provided in the central lab.

And, what is important, new technologies improve the quality and the validity of the results. The rapid development of POCT is in the agreement with current trends in the healthcare and POCT could be the main diagnostics tool in 21st. century.

We have to know where are the main sources of errors for the assurance of quality patient care, including. The errors can arise in different parts of the whole process – inappropriate tests, inappropriate evaluation of results, errors in the analytical and the preanalytical process etc. We can evaluate errors when we judge the whole process on the principle:

The right result, for the right investigation on the right specimen from the right patient, available at the right time, interpreted using the right reference data, and produced at the right cost.

POCT reduces the risk of errors in the pre-examination and the post-examination processes. Additional, current POCT technologies are guarantees valid results.

Of course, quality assurance system depends on the location and on the type of POCT. If is POCT in hospital, quality assurance is in the hands of laboratory staff. The quality assurance for the other types and systems (GP offices, home, off-site...) depends on the keeping of the rules and the regulations.

The most effective management for ensuring quality in any situation is a quality system (3, 4). Various quality systems are in use, though there is an increasing of the ISO standards. The quality system has a number of components. All components of a quality system should be in place and operating in order to fully achieve the end product of good quality laboratory service. Excessive attention to any one individual component, to the neglect of others, will not achieve lasting improvements in quality. A balance approach is essential.

2. POCT – Current situation

The 2004 tsunami in Southeast Asia and Hurricane Kathrina in the USA revealed significant weakness in public health planning, immediate rescue and follow-up care (1, 2). These disasters also convinced political and health care leaders of the need for rapid response and readiness. Rapid development of POCT technologies parallels current trends towards distributed health care. Point-of care testing is well positioned to become a dominant diagnostic approach in the 21st century.

The professional point-of care market, as distinct from the very large home testing area, is composed of 2 general segments: hospital testing and decentralized testing. Hospital POCT is usually an extension of central laboratory testing. For example, immediate turnaround of blood gases and electrolytes are very helpful in the operating room or the emergency department. Pregnancy tests or cardiac monitoring tests in the emergency department can be very helpful to manage patients with critical illnesses. However, here POCT can be used in distinctly different way to provide decentralized testing of selected parameters on whole blood in key locations throughout the hospital.

A main component of decentralized testing segment consists of physician office laboratories, nursing homes, pharmacies, and other noninstitutional settings in which health care providers perform fast, simple near patient diagnostic tests for the purpose of screening, therapy, and disease monitoring. Key physician groups here include general practitioners, cardiologists and internal medicine specialists.

Is very important to know what tests are provided by POCT, what tests are important to be a part of the POCT and what techniques are used in POCT (15). We can find very different requirements for the POCT tests. It usually depends on the author, on his specialization. Different view exists from laboratory specialists, from hospital clinicians and from practitioners. In table 1 you can find variety tests available in the POCT format for some group of tests. However, here are only the main tests here.

Group	Tests	material
Diabetes	HbA1c, glucose	blood
	Microalbumin	urine
Drug of Abuse screening	The whole spectrum of drugs	urine, saliva, sweat
Alcohol	Alcohol (Ethanol)	saliva, blood
Metabolites	Lactate	blood
Lipids	Cholesterol, TAG, HDL	blood
Pregnancy, hormones	hCG, LH, FSH, PTH	blood, urine
Urine strips	Chemistry	urine
Coagulation	PT, APTT	blood
Renal tests	Creatinine, Urea	blood
Liver tests	ALT, GGT	blood
Cardiac markers	TnI, TnT, Myoglobin, BNP...	blood
Serology	HIV, helicobacter pylori etc	blood, saliva
Hematology	Hemoglobin, hematocrit	blood

Table 1. Group of tests in POCT

Devices for POCT must be simple to use, robust both in terms of storage and usage, capable of producing results consistent with the clinical requirement, and safe both in storage, use and disposable(12, 15 ,16). Analytes that require only a qualitative response can be detected with stand-alone disposable devices that do not need instrumentation, although there have been methods where a quantitative result can be achieved without instrumentation. However, in the majority of cases when a quantitative result is required, the disposable analytical device requires a reader. The reader may also facilitate or control parts of analytical reaction, e.g., temperature control, timing of absorbance reading, fluorescence, etc. The design template for any POCT device is based on the creation of the cell in which the reaction takes place. This cell can be either a porous material, a chamber, or a surface.

We can divide POCT devices to the three main categories – single-use devices, hand-hold sensor systems and bench top small instruments. Of course, many systems could be the combination of above mentioned or untypical. Single use devices are the oldest. The typical example is urine strip. However, various detection systems can be used for single-use disposable devices. The simple classification is in the Table 2

Technology	Detection
Single or multi-pad stick	visual read, reflectance read
Enzyme chromatography	analog signal
Immunoassay flow through or lateral flow	visual read, reflectance read
Immunoassay chromatography	analog signal
Cassette devices	optical read
Microfabricated devices	optical read
Biosensors	electrochemical, optical diffraction, ellipsometry, etc.

Table 2. The main technology in POCT

The development of microsensors opened in the late 90´s way for so called hand hold sensor systems. These systems are easy portable with better analytical quality than previous technology. A few analytical principles are using for hands-hold devices: enzyme based systems (optical or amperometric detection), optical motion detection (for coagulation with magnet detector), ion-selective electrodes (amperometric or potentiometric detection), immunochemistry (visual, optical detection) and solid phased chemistry (reflectance detection).

The last systems (bench top) are near to the classical laboratory devices. There are very similar detection systems, but usually it working with the whole blood.

3. The development of POCT

The scope of POCT now includes testing in home, pharmacy, supermarket, physician´s office, and at hospital bedside. Future expansion in the scope and volume of testing may depends on a series of factors. These include government healthcare policy and laws about POCT; the availability of POCT; the view of general public on POCT; cost of testing; new technologies that make POCT easier, more reliable, and more likely to be accepted; and broader developments in Web-based electronic health record.

3.1 Health care policy, regulations, and availability of POCT

Sale of tests for use by members of the general public or direct access of the general public to a testing service is regulated in many countries (17). In the US, for example, clinical testing is regulated by the Clinical Laboratory Improvement Amendments (CLIA) which was passed in 1998. A test must achieve a waived test status for it to be available to the general public for use as a self-test. An alternative source of POCT is to purchase testing at one of a number of retail health clinics that have sprung up in supermarkets and pharmacies. Although above mentioned tests offer a new source of POCT for the general public, they have met with opposition from medical professional organization. Variety of recommendations and guidelines exist for POCT. Finally, it is possible for the general public to purchase a sampling kit and send in a specimen to a laboratory and have a variety of clinical tests performed – so-called direct access testing.

3.2 New technologies for POCT

An important factor in the future of POCT is the technological innovation that makes it easier, more reliable, and more likely to be accepted (1, 12, 16). Such developments provide a possible and significant stimulus to POCT that has already seen an amazing progression from the early tablet-based tests do dipsticks, to quantitative measurement using meters, then to all-in-one 1-step-devices, and finally, to real time assays using implantable sensors.

The next source of innovation is being drawn from a number of areas. These include the consumer electronic industry, and examples include the development of disposable POCT devices that include electronic components, such as some pregnancy or HbA1c tests. In addition, developments in Wi-Fi and graphical interfaces promise connectivity and the ease of use now associated with cell phones. Another source of technology adaptable to POCT includes the extensive developments in devices for field-testing for bio-warfare agents.

These range from chemical agent monitors to integrated polymerase chain reaction-based devices for infectious agents. The design specifications for a device for field-testing for bio-warfare agents and a POCT device for clinical testing share many similarities – inexpensive, quick, accurate, robust, easy to use, and connected – and the sizable research and development investment in this area should eventually provide spin-off into POCT.

Two technology areas that hold promise for POCT are microtechnology and nanotechnology, principally because they offer routes to miniaturization and simplification of POCT devices. Microtechnology has already provided a wide range of miniaturized analyzers (so-called lab-on-chip type devices) that in principle have a potential for POCT. In particular, microtechnology enables integration and miniaturization of complex multistep and dynamic tests, thus, extending the range of tests that could migrate to POCT, such as polymerase chain reaction – based assays and flow cytometry. Nanotechnology offers an exciting array of new analytical strategies and nano-sized structures that have analytical potential. Typical examples are a DNA sequencer based on a nanopore fabricated in a small chip and contact lens-based glucose analyzer that use nanoarrays of beads or gratings to detect glucose in fluid bathing the eye.

3.3 Web-based electronic health records

There has been an increased interest in providing Web-based locations for collecting and storing medical information. New electronic health care aids are intended to be safe and secure and to help patients gather medical records from physicians, hospitals and pharmacies. Electronic systems are also designed to connect with personal health and fitness devices to consolidate all medical information in one place. A possible consequence of this new direction for health care is that patients will become more interested in gathering clinical data and thus, become more motivated to perform self-testing to obtain the data they seek. The net result would be an expansion in POCT.

4. Advantages and disadvantages of POCT

4.1 Advantages

For POCT, the analytical time is short, on the order of minutes, allowing for a result often while the patient is still being examined. This decreases the therapeutic time and offers the "potential" to improve patient outcome by reducing delays in wait time for laboratory results. When patient management is dependent on the test result, laboratory testing can be a significant bottleneck, resulting in longer wait times and poorer outcomes in particularly unstable patients. POCT, providing a rapid result, can prevent such bottleneck.

Clinician time can also be reduced, since physicians can obtain results while the patient and diagnostic issues are still fresh in his minds. If a sample needs to be collected and sent to a laboratory, the clinician is going to move on to other patients. Once test results are available, clinicians will need time to refamiliarize with the patient, their diagnosis, and where they left off in the treatment pathway. Thus, POCT has the potential to improve physician efficiency by enhancing result turnaround time and better fitting within clinical workflow.

POCT also reduces the risk of errors in the pre-examination and the post-examination processes. Above all are reduced pre-examination errors, there are no confusion of patients (the same exist for post-examination errors), no errors in sample transport.

The small volume of specimen required for POCT is also benefit for patients with bleeding disorders, neonates, and other for whom collection of standard sample volume could be a problem. POCT utilizes unprocessed specimens, which further reduces the amount of specimen required for tests such as electrolytes, glucose, and enzymes that are traditionally conducted on serum or plasma in centralized laboratories, requiring larger volume for centrifugation and the removal of erythrocytes. Elimination of sample processing is an additional feature of POCT that contributes to a faster result.

4.2 Disadvantages

Managing the quality of POCT is a challenge due to the number of different devices and scenarios for delivery of testing. With the rapid rise in POCT popularity, quality has become a growing concern. There are many different problems – result´s misleading, bad storing of reagents, poorly cleaned POCT devices (reservoir of nosocomial). Misleading results have been noted on proficiency surveys, comparative that are sent blinded to all laboratories performing testing in order to determine the accuracy of the testing process. Higher failure rates have been noted in POCT when compared with hospitals, with the same testing devices. Historically, the better performance has been linked to staff competency and having staff trained in laboratory science supervising the testing process.

Staff training and clinical focus may be one source of quality problems. Clinical staff does not have formal laboratory education and tend to be patient centered. The performance of laboratory testing requires dedicated attention to detailed, stepwise analysis. When a patient is in trouble, clinicians and nurses are going to react in the aid of the patient. The laboratory test becomes a secondary concern. Staff may forget to precisely time the testing device, or may not take note of specific sample collection details. Nurses and clinical staff may also not fully appreciate the preanalytical variables that can interfere with the test result. They often assume that if the test gives a result, it must be the correct result. This leads staff to believe that POCT is equivalent and freely interchangeable with a central laboratory test. The glucose is a glucose test, a pregnancy is a pregnancy test, etc. Historical management pathways developed from the use of central laboratory tests now utilize POCT without necessarily accounting for differences in methodology or test limitations. However, POCT differs still in precision and accuracy from its central laboratory counterpart. It utilizes different chemistries in order to provide a faster result. Thus, POCT has technical unique from the central laboratory methods.

Patient population characteristics can also contribute to POCT and central laboratory differences. Hospital use of POCT stresses the technical performance of devices that are approved and marketed for home use applications. Home patients are ambulant, generally well, and utilize easily collected samples, such as urine, saliva or capillary blood. Yet, hospitalized patients have a variety of acute and chronic illnesses, are confined to bed, and are on number of medications. Use of device in this patient population may be challenged by unusual samples, such as arterial and venous blood and blood drawn from intravenous line. Hematocrits are higher (neonates) or lower (oncology, surgical and trauma patients) than in the outpatient population. POCT on inpatients is thus a very different scenario. Application of devices to patient populations that have not been validated through clinical trials may lead to result discrepancies that were not predicted from studies on home use patients.

5. POCT Quality management

Quality management is the total process of supporting the quality of the testing service and is comprised of measure taken to ensure investigation reliability. These measures start with the selection of appropriate tests and continue with the obtaining of a satisfactory sample from the right patient, followed by accurate and precise analysis, prompt and correct recording of the result with an appropriate interpretation, and subsequent action on the result. Adequate documentation forms the basis of a quality assurance system in achieving standardization of methods and traceability of results on individual specimen. Regular equipment monitoring, preventive maintenance and repair when needed are other important components of quality management.

5.1 Standards and accreditation

Accreditation of laboratories or of users of POCT is a process of inspection by a third party to ensure conformance to certain pre-defined criteria. The main sources for accreditation are the ISO standards – for laboratory ISO EN 15189:2007 - Medical laboratories – Particular requirements for quality and competence. The special standard exists for POCT – ISO EN 22870:2006: Point-of- care testing (POCT) – Requirements for quality and competence. This special standard includes only additional requirements to the 15189 (10, 13).

In the US, under the CLIA all clinical laboratory testing including POCT examining "materials derived from the human body for the purpose of providing information for the diagnosis, prevention, or treatment of any disease..." is regulated. These CLIA mandates are based on test complexity – waived and nonwaived – and focus on 3 phases of testing: preanalytical, analytical and postanalytical. Does it mean all testing procedures performed POCT must be conducted under an appropriate CLIA certificate. Point-of-care testing sites within a hospital typically fall under 1 or 2 broad scenarios: (1) central laboratory holds a single CLIA certificate that covers all institutional testing, including POCT, or (2) POCT sites within the institution have their own, separate CLIA certificates. Freestanding POCT sites, such as physician office laboratories (POLs) and clinics, must have their own CLIA certificates. CLIA includes requirements for procedure manuals, quality systems, method verification, quality control, personnel, proficiency testing and inspection. Voluntary accreditations also exist in US – the Joint Commission(special standards for POCT), the laboratory accreditation program of the College of American Pathologist and the Commission of Office Laboratory Accreditation (COLA).

5.2 Standardization and harmonization

Standardization is a practical process aimed at achieving consistency among measurement procedures through application of high scientific standards (4, 12). Standardization has become an important concept in clinical diagnostics, both nationally and globally. Key components of standardization include higher-order materials and measurement procedures, such as the Standard Reference Materials (SRMs) for diagnostic tests developed by the National Institute of Standards and Technologies (NIST).

Results of routine daily measurement are standardized through calibration to the reference method and/or material (traceability) but, when the reference system is lacking, are only referred to a manufacturer's selected procedure and corresponding calibrator. Ideally, for example, manufacturers of metabolite assays (eg. glucose) could use SRM 1950, once fully developed, to assess the validity of their calibrators. This practice will result in standardized measurement with closer agreement of results in clinical practice. In support of the standardization practice, societies such as the International Federation of Clinical Chemistry and Laboratory Medicine (IFCC) have set forth to promote extensive global standardization in laboratory medicine.

The POCT motion of bringing the test immediately to the patient is accomplished via convenient handheld, portable, or transportable devices, many of which used whole-blood sample for result. The rapid technological advances in POCT permit measurement of multiple analytes in whole-blood samples. However, the lack of whole-blood SRMs reveals the standardization deficiency in POCT field that can adversely impact diagnoses, treatment decision, and patient outcomes. Therefore, the first is for global standardization in POCT, which will ultimately result in an improvement of clinical interpretation of laboratory results to benefit patient.

Harmonization in medicine means bringing diagnostic standards and treatments into consonance or accord. However, at this point in time, no single reference or calibration method correlating plasma and/or whole-blood glucose and other analytes has been adopted universally among manufactures. Additionally, if we adopt a bedside perspective, then most reference instruments should not show statistically significant bias relative to the point of care. Currently, however, there is a proven lack of harmonization of reference instrument used in hospital across the US or Europe when viewed from bedside perspective. This deficiency in harmonization potentially introduces complications in clinical management, which can adversely affect treatment and care of patient.

Now, adding "global" into the equation would mean unifying international standards to create worldwide harmony in diagnostic testing. In the absence of global harmonization, the likelihood that a result may be misinterpreted and the patient may be misdiagnosed is increased because of different assays exhibiting nonequivalent analytical responses. A spectrum of different values may result from testing of the same specimen by different methods. Data reporting comparisons of global SRMs show an international bias resulting in noncomparable patient outcomes due to different field methods. Bias in measurement has also shown to impact medical decision making.

Additionally, too long, the perspective has been clouded. Preanalytical delays and other confounding factors deteriorate samples before they are assayed in clinical laboratories. In many cases, bedside results may reflect actual in vivo condition more accurately.

5.3 Quality Assurance

Quality Assurance (QA) is a management approach that attempts to ensure the attainment of quality, enabling appropriate and timely clinical action (11, 12, 13). QA thus encompasses much more then analytical quality. POCT presents particular quality concerns. Surveys have shown that POCT, carried out by non-laboratory staff outside conventional laboratories, is widespread in both the hospital sector and the community sector, with an increasing range of tests and number of testing sites. Systems used for POCT are increasingly sophisticated, with more aspects of quality built in or controlled through manufacturing quality control. The equipment and procedures seem easy to use, but quality is not provided automatically and QA is essential. The first aspect of quality in POCT is whether the test should be done at all. Before such testing is introduced or extended, clinicians must discuss and agree on the pattern for service provision, based on clinical needs. Even where investigations are deemed appropriate for POCT, audit of testing quality and effectiveness should be ongoing.

The most effective management tool for ensuring quality in any situation is a quality system. Various quality systems are in use, though there is increasing utilization of the International Organization for Standardization's (ISO) 9000 series of international standards. Accreditation and quality systems are in principle as applicable to POCT as to conventional laboratories, though practice and legislation may differ from country to country.

All components of a quality system should be in place and operating in order to fully achieve the end product of good quality laboratory services. Excessive attention to any one individual component, to the neglect of others, will not achieve lasting improvements in quality. A balanced approach is essential.

Components of Quality system:

Quality Assurance. QA is the total process of supporting the quality of the testing service and is comprised of measures taken to ensure investigation reliability. These measures start with the selection of appropriate tests and continue with the obtaining of a satisfactory sample from the right patient, followed by accurate and precise analysis, prompt and correct recording of the result with an appropriate interpretation, and subsequent action on the result. Adequate documentation forms the basis of a quality assurance system in achieving standardization of methods and traceability of results on individual specimens. Regular equipment monitoring, preventive maintenance, and repair when needed are other important components of QA. Quality assurance must not to be limited to technical procedures. Those who obtain specimens for analysis make a significant contribution to the reliability of the results through correct specimen collection and handling. Inappropriate specimen collection is a major source of variation, which must be minimized through careful training and adequate supervision.

Internal Quality Control (IQC). IQC assesses, in real time, whether the performance of an individual testing site is sufficiently similar to its previous performance for results to be issued. Thus IQC controls reproducibility or precision, ensuring that sequential results are comparable and credible, and maintaining continuity of patient care. Most IQC procedures analyze a defined control material and ascertain if the results obtained were within previously established limits of acceptability.

External Quality Assessment (EQA). EQA, in contrast to IQC, compares the performance of different testing sites. This is made possible by the analysis of an identical specimen at many sites, followed by the comparison of individual results with those of other sites and the "correct" answer. The process is necessarily retrospective, and it provides an assessment of performance rather than a true control for each test performed on patients' specimen.

Audit. Audit is a process of critical review. Internal audit examines local processes conducted by senior staff. Such reviews measure various parameters of performance, such as timeliness, accuracy and cost of reports, and identification of weak points in the system where errors can occur. External audit widens the input by involving others in the evaluation of analytical services. The users of services are asked how they perceive the quality and relevance of the service provided.

Accreditation. Accreditation of laboratories is a process of inspection by a third party to ensure conformance to certain pre-defined criteria. Accreditation can be linked to a formal system of licensing, whereby only accredited testing sites are legally entitled to practice, or it may be a voluntary system. In many cases, accreditation of POCT sites is subsidiary to the accreditation of the laboratory responsible for them, through independent accreditation of POCT site may be permitted in some circumstances.

Validation of Results. Validation is an attempt to measure quality by re-examining previously analyzed specimen.

Training and Education. This is probably single most important component in any QA program. Issues to be addressed include national policies and curricula for staff training, both during primary training and in follow-up. Professional status and career development are related factors. Training should be continually monitored, and where new needs arise (such as introduction of new methods), or where QA reveals the need for improvements, courses and workshops may be introduced.

Evaluation. Determining whether reagents and equipment are suitable for use may make significant contribution to overall quality. Choice of equipment and reagents requires considerable thought and co-ordination and may be cost effective in reducing duplication of effort and preventing repetition of expensive mistakes. It cannot be assumed that equipment and assays designed for use in one situation will perform well in another.

Documentation. Documentation of all procedures is a universal accreditation requirement, and incorporation of controlled documents within the framework of a formal quality manual in accordance with ISO 9000 standards is highly desirable.

5.4 Internal Quality Control

IQC in laboratory medicine was developed by adapting systems used in the manufacturing industry. In laboratory practice, however, specimens are not expected to be identical, so quality control material must be included with each batch of patients' specimen to probe the performance of analytical system (5, 12, 14). Validated procedures have been developed for the statistical interpretation of control data generated from quantitative analysis, though these may not be directly transferable to establishments carrying out POCT. Some systems used in decentralized testing either do not require any calibration (standardization) by the user as they are factory-calibrated or require infrequent recalibration. Much of the variability of results from these methods originates from variations in operator technique. The analysis of an appropriate control material before starting and during analysis of clinical specimens can provide reassurance, that the system and operator are performing correctly.

Results obtained with control materials must be recorded and interpreted. Representation in graphical form is strongly recommended. For a quantitative analysis, control results must be compared with the acceptance limits determined as from previous experience as on the basis of analytical goal settings. There is an internationally accepted, 5-tiered hierarchical model for this setting of analytical goals for the imprecision, bias, and total allowable error of laboratory tests. Where data are available from more than 1 approach, models higher in hierarchy are considered to hold greater weighting than those from lower levels. The highest quality standard is required when analytical quality has a direct effect on medical decision making in a specific clinical situation (e.g. HbA1c). Analytical goals for broader clinical need can be derived from biological variability or from clinical survey how clinicians use test results. Three classes of analytical goals (minimum, desirable and optimal), based on fraction of within-individual biological variation, have also been developed for the imprecision of commonly tests. The desirable analytical goal for most biochemical analytes is that the analytical imprecision (Coefficient of variation, CVa) should be less than one half of the average within-person biological variation (0,5 CVw). However, for those analytes for which the desirable goals derived using this formula are readily achievable with current methodology, it is recommended that an optimal analytical goal be used, based on the formula CVa ≤ 0,25 CVw. For those analytes for which desirable goals are not readily attainable by current methodology, a minimum analytical goal is recommended, based on formula CVa ≤ 0,75 CVw. Many national and international groups have also set profession-defined analytical goals; government or external quality assurance program organizers also set analytical quality specification.

As an overarching principle, analytical goals for POCT should be equivalent to those used for laboratories to ensure that the use of POCT does not compromise standard of patient care and clinical decision making. However, it is important to acknowledge that the POCT environment is often very different to the laboratory settings. First, long-term retention of staff as POCT operators is an ever- present problem for rural and remote health services, creating difficulties in sustaining not only POCT but other health programs in general. Second, there needs to be a balanced approach to goal setting. There is limited value in setting analytical goals there are too stringent and which current POCT instruments cannot achieve. On the other hand, there is a clinical imperative to ensure that excessive analytical noise from POCT methods does not mask clinically significant changes in patient results. Minimum analytical goals for the imprecision of non-laboratory based POCT should be set where sufficient data are available with the proviso that performance outside the minimum goal shoul be investigated and acted upon by the clinician responsible for clinical governance of POCT. Third, when assessing analytical performance in POCT environments again published literature, should be noted that the majority of published evaluation of POCT instruments have been conducted in the laboratory settings by trained laboratory staff. Analytical goals may also vary depending upon the intended purpose of the test. For example, different goals may be needed for diagnosis versus monitoring of test result.

The main problem for IQC in POCT is in personal working with it. These workers have no education in the laboratory field, poor knowledge about QC in laboratories. So, the IQC system must be easy and simple. IQC must be run with appropriate frequency. Ideally, IQC should be run with every batch of clinical specimens, but this may not be practicable or necessary. In POCT, specimens are often analyzed individually rather than in batches, and running IQC once per operator per shift (before analyzing the first clinical specimen) appears satisfactory in most situations.

The main requirements for IQC in POCT:

- Controls for blood gases and electrolytes measure on three levels each day. If are used instruments with equivalent QC, the QC system must be validated before using the instrument for patient. Following the producer rule is the basic requirement.
- Glucometres must be daily control on two levels of control material
- Personal glucometres must be regularly controlled by some lab.
- The other instrument and test must be controlled daily
- Evaluation of results IQC is on the same rules and processes as in the labs.

The best approach to IQC for POCT is keeping of standard procedures by laboratory staff. But it isn't usually assured, many instruments and techniques are outside of some hospital or laboratory. Some instruments are so called single use devices. From these reasons exists equivalent system IQC:

- Instruments with fixed controls inside and with regular measuring (7),
- Equivalent controls (8),
- Electronic control (EQC),
- Automatic controls (onboard control, OBC),
- iQM (intelligent quality management) (6)

Equivalent controls (Eqc) is the procedure allows further reduction from daily QC to weekly or monthly QC for certain analytical systems that contain internal procedural controls. To implement eqC, the test site needs to determine the extent to which the manufacturer's internal/procedural controls adequately evaluate and monitor the analytical process and test components. The test site then selects the appropriate eqC option for evaluation. There are 3 options – option 1 is for test systems with internal/procedural control(s) that monitor the entire analytical process; option 2 is for systems that monitor a portion of the analytical process; option 3 is for instruments that monitor none of the analytical process.

For single-use devices without an instrument for reading the results, internal procedural controls may have been built-in by manufacturer. These are designed to ensure that results can only be obtained if the specimen has been applied properly and the reagents have worked correctly, and are particularly useful for non-quantitative. i.e., positive or negative, investigation. Nevertheless, acceptance testing of each lot or container of devices with positive and negative controls may provide an invaluable quality control measure before issue to testing site.

Some systems are equipped with electronic control. Here a "dummy" device module mimics electronically the output from a real device, so that the reading instrument gives a valid "result"; the same module may mimics several different concentrations. These modules are stable, and provide excellent validation that the reading instrument is working correctly for maintenance or troubleshooting purposes. The EQC unit can be use repeatedly to check the reader portion without having to reconstitute control solution or use the disposable portion of the test system. EQC is not new. Photometer checks, cuvette checks, blank (zero) checks, and a variety of electronic voltage checks have long been used as a part of the monitoring system for laboratory instrument.

As unit-use test systems evolved, many manufacturers began to introduce products with a control that was included as an integral part of the disposable component of the device. This is often called automatic control or onboard control (OBC). With this method, some type of control is built into the disposable component and provides a QC result that is run simultaneously with the patient sample. Current versions include calibration checks and multiple channels for low and high controls; in one system, three levels of controls are run.

Intelligent quality management (iQM) is proposed on IL's GEM Premier 3000 as an alternative to traditional IQC. The theoretical aspects of this concept have been evaluated by Westgard, who has demonstrated the potential value of an increase in the frequency of control testing.

5.5 External Quality Assessment

EQA, in opposite to IQC, compares the performance of different testing sites (9). This is made possible by the analysis of an identical specimen at many sites, followed by the comparison of individual results with those of other sites and the „correct" answer. EQA is retrospective, and it provides an assessment performance rather than a true control for each test. An EQA, or proficiency testing program, when applied to POCT, is designed to provide regular, objective and independent assessment of an in vitro medical devices and its operator´s ability to provide an acceptable standard of service by comparison with peers. POCT EQA is equivalent to the well established processes of interlaboratory comparison in the central laboratory setting. It is an integral part of the total quality management system. The process is organized by the impartial agency.

5.6 Training and education

This is probably the most important component in any quality assurance program. Issues to be addressed include national policies and curicula for staff training, both during primary training and it follow up. Professional status and career development are related factors. Training needs should be continually monitored, and where are needs arise, or where are QA reveals the need for improvements, courses and workshop may be introduced.

Errors are more often associated with lack of understanding, inadequate training, and miscommunication rather than the analytical systems. Appropriate training of both the trainer and the user is therefore essential for a first class POCT service. The management of the training and competency assessment is usually managed by POCT coordinator, more often recruited from within the laboratory rather than from a nursing background. Often the training requirements vary greatly between institutions. Both theoretical and practical training program for all POCT personnel should be implemented where only personnel who have completed training and demonstrated competence shall carry out POCT. The knowledge/skill requirements include the ability to demonstrate an understanding of the appropriate use of the device, the theory of the measurement system, and appreciation of the preanalytical aspects of the analysis, including the following:

- sample collection,
- its clinical use and limitations,
- expertise in analytical procedure,
- reagent storage,
- QC and quality assurance,
- Technical limitation of device,
- Infection control practice,
- Response to result that fall outside of predefined limits,
- Basic maintenance of the device,
- Correct documentation.

Operators´ competence should be objectively and independently assessed through practical demonstration. Records of training/attestation and of retraining should be retained as evidence.

6. Conclusion

The quality in the POCT is on the way for improvement, but in the comparison with laboratories is still not so good. Sigma value is usually below 3.5, standardization and traceability are poor, and systems of IQC are in the beginning. Fortunately national recommendations are arising, some strict rules also exist (ADA, Rilllibak, CLIA). However POCT systems will certainly develop and quality assurance will be better and better.

7. References

[1] Kricka L. J. Point-of-Care Technologies for the Future, *Point of Care*, Vol. 8, No.2, (June 2009), pp. 42-44, ISSN 1533-029X

[2] Dooley J.F. Point-of-Care Diagnostic Testing Markets, *Point of Care*, Vol. 8, No.4, (December 2009), pp. 154-156, ISSN 1533-029X

[3] Ehrmeyer S. S. The US Regulatory Requirements for Pont-of-Care Testing, *Point of Care*, Vol. 10, No.2, (June, 2011), pp. 59-62, ISSN 1533-029X

[4] Gentile N. L., Louie R. F., Mecozzi D., Hale K., Kost J.G. Standardization, Harmonization, and Realization, *Point of Care*, Vol. 7, No.3, (September, 2008), pp. 110-112, ISSN 1533-029X

[5] Shephard M. D. S. Analytical Goals for Point-of-Care Testing, *Point of Care*, Vol. 5, No.4, (December, 2006), pp. 177-185, ISSN 1533-029X

[6] Vaubourdolle M., Braconier F., Daunizeau A iQM: A New Concept for Quality Control, *Point of Care*, Vol. 3, No.3, (September, 2004), pp. 177-185, ISSN 1533-029X

[7] Phillips D. L. Quality Control for Unit-Use test systems, *Point of Care*, Vol. 4, No.1, (March, 2005), pp. 58-60, ISSN 1533-029X

[8] Ehrmeyer S. S., Laessig R. H., Darcy T. P. CLIA´s Concept of Equivalent Quality Control, *Point of Care*, Vol. 4, No.2, (June, 2005), pp. 177-185, ISSN 1533-029X

[9] Richardson H., Gun-Munro J.,Petersons A., Webb S. External Quality Asessment in the Quality Mangement of POCT, *Point of Care*, Vol. 4, No.4, (December, 2005), pp. 177-185, ISSN 1533-029X

[10] Thomas M.A. Quality Assurance and Accreditation in Point-of Care Testing, *Point of Care*, Vol. 7, No.4, (December, 2009), pp. 227-232, ISSN 1533-029X

[11] Fraser C. G. General strategies to set quality specifications for reliability performance characteristic, *Scan. J. Clin. Lab. Invest.* Vol 59 (1999), pp 487-490

[12] Price Ch. P., Hicks J. M. (1999) *Point-of Care Testing*, AACC Press, ISBN 1-890833-23-9

[13] Westgard J. O., Ehrmeyer S. S., Darcy T. P. (2004) *CLIA Final Rules for Quality Systems*,Westgard QC Inc., ISBN 1-886958-20-3

[14] Fraser C. G. (2001) *Biological Variation, from Principle to Practice*, AACC Press, ISBN 1-890883-49-2

[15] Kost G. J. (Ed(s).). (2002) *Principles & Practice of Point-of-Care Testing*, Lippincot Wiliams & Williams, ISBN 0-7817-3516-9

[16] Nichols J. H. (Ed(s).). (2003) *Point-of-Care Testing, Performance, Improvement and Evidence-Based Outcomes*, Marce Decker Inc, ISBN 0-8247-0868-7

[17] Price Ch. P., St John A. (2006) *Point-of Care Testing for Managers and Policymakers*, AACC Press, ISBN 1-59425-051-0

6

Quality Management and Medical Education in Saudi Arabia

Ali M. Al-Shehri

King Saud bin Abdulaziz University for Health Sciences, National Guard Health Affairs
Kingdom of Saudi Arabia

1. Introduction

More than 100 years ago Abraham Flexner's report changed medical education in the United States (USA) and consequently worldwide. He simply used a gold standard against which all medical schools at that time were measured (Cook et al., 2010). The report at that time was concerned mainly with undergraduate medical education. Since then medical education has developed immensely and flourished into many branches and specialties, becoming a continuum: undergraduate medical education, postgraduate medical education (training and residency programmes) and continuing medical education. Clearly both the quantity and quality of medical education have undergone major improvements over the intervening 100 years, but still there are quality issues at all levels of medical education recognized in both developed and developing countries that need to be addressed (Tekian & Al-Mazroa, 2011; Cook et al., 2010; Hasan, 2010; Ehlers, 2099; Took, 2008; Sood & Adkoli, 2000; Maddison, 1987; Lowry, 1993; Al-Shehri et al., 2008, 1993, 1994; Al-Shehri & Al-Gamdi 1999; Al-Shehri, 1996, 1993, 1992).

For undergraduate medical education, it has been reported that "medical colleges are producing graduates who are not well equipped to tackle the health care needs of the society" (Sood & Alkoli, 2000). The gap between theory and practice, and the failure of medical education to foster critical thinking and self-directed learning among medical students are recognized in both developed and developing countries (Lowry, 1993; Al-Shehri, 1999). For postgraduate medical education, a number of issues have been identified: unclear policy for postgraduate medical training; lack of consensus on the role of doctors at different stages of their career; fragmented and weak governance, and generally medical training is "unlikely to encourage or reward striving for excellence" (Tooke, 2008). For continuing medical education there is no evidence that current provision is subjected to quality management that takes into consideration structure, process and outcome (Al-Shehri et al., 2008). So the issues of medical education through its continuum are no longer quantity, but rather quality, which is more prevalent in developing countries.

Quality management transformed Japan's manufacturing industry to become a gold standard for developed countries. Judging from the success of quality management in the manufacturing industry, one may argue that similar quality management in medical education is very much needed for a number of reasons. These include ensuring that products meet customer needs; assisting in raising people's level of health; affirming quality of teaching and learning on the basis of sound evidence; enhancing public confidence in the educational system; prestigious and ranked recognition of academic organizations; financial and academic attractions and global market competitiveness (Cueto et al., 2006; Woodhouse, 2003; Gale & Grant, 2010). Despite these legitimate reasons, one wonders if quality management as defined and used by the industry is acceptable to all stakeholders of medical education. It has been reported that quality management is perceived by academic staff in medical schools as "overly onerous and detrimental to their real work" (Hasan, 2010).

Quality management depends largely on customers' satisfaction (Berwick, 1992). In the manufacturing industry, customers are not usually part of the process of production while in medical education and health care they are an integral part of it (Gale & Grant, 2010; Berwick, 1992). Many intrinsic and extrinsic factors interplay to influence structure, process and outcomes of medical education, and make it more complex than the manufacturing industry. For example, it has been reported that "the administrative structure of the traditional medical school is designed to achieve maximum inflexibility and the greatest possible difficulty in adapting the school's policy and programs to a rapidly changing world" (Maddison, 1978). Moreover, in medical education different stakeholders (customers) may have different, if not contradictory, needs and expectations (Al-Shehri, 2003). Although it is unconceivable to adopt quality management that does not consider the complexity of medical education, the most commonly used approach to quality management in medical education in both developed and developing countries, is accreditation.

1.1 Accreditation of medical education

Accreditation has been defined by the European Consortium of Accreditation (ECA) as "a formal and independent decision, indicating that an institution of higher education and/or programs offered meet certain standards" (ECA, 2005).

Woodhouse (2003) counts more than 140 bodies of quality accreditation and certification in relation to higher education. Each accrediting body has its own way of definition. The danger here is that unless stakeholders of medical education believe in quality as a culture, seeking accreditation will become a superficial exercise that depends on filling in boxes which may or may not reflect the actual beliefs, attitudes and behaviours of medical education organizations. Moreover, the multiple factors influencing the learning-teaching process cannot be addressed completely by assessment and accreditation. "While the quality of teaching and learning interaction between students and educators in higher education is influenced by a variety of factors, including attitudes and skills of teachers, abilities and motivation of learners, organizational backgrounds, contexts and values and the existing structures, such as rules, regulations, legislation and so on, the majority of approaches to assess, assure, manage or develop quality is only partially taking these factors into account" (Ehlers, 2009).

Differences in culture between developed and developing countries may also influence the perception and purposes of accreditation. Political, social and cultural factors in developing countries may equate accreditation with quality achieved; a destination rather than a journey and an outcome rather than a process. Failing accreditation, on the other hand, may have drastic consequences on the development of medical education particularly in developing countries where there is always a need to expand medical education to cope with the growing needs of the society and the development of the country. Although accreditation may not be the best choice for developing countries, it is spreading.

In their analysis of nine developing countries, Cueto et al. concluded that "the trend towards instituting quality assurance mechanisms in medical education is spreading to some developing countries" (Cueto et al., 2006). Four main steps must be followed for accreditation:

- Self-study report and submission to the accrediting body;
- Formal survey visit by representatives of the accrediting body;
- Recommendation of the surveying visitors;
- Accreditation decision by the highest governance of the accrediting body.

Accreditation decisions can be full accreditation, provisional accreditation, no accreditation or withdrawal. Duration of accreditation lasts usually from five to eight years before the organization is due for reaccreditation (Cueto et al., 2006).

Unlike developed countries, accrediting bodies of medical education in most, if not all, developing countries are governmental agencies under ministries of education and health. This may raise certain issues related to bias, control, validity and reliability of the outcome of the accreditation process. Passing or failing accreditation has certain connotations politically, professionally, organizationally and financially. To put the accrediting bodies under the control of a ministry of education and health may lead to either collusion or conflict of interests that cloud the interpretation of results. Regardless of this dilemma of collusion and conflict there is no strong evidence to equate accreditation with high quality outcomes of medical education, particularly so in developing countries.

Saudi Arabia, the largest country in the Gulf region, has witnessed unprecedented expansion in higher education. This includes more than 100,000 students sent abroad mostly in health related specialties; medical schools increased by 200% over the last five years with huge intakes; more than 60 postgraduate medical training programmes and fellowships run annually; provision of e-learning programmes and thousands of continuing medical education activities and programmes provided annually (Al-Shehri, 2010; Al-Shehri et al., 2008, 2001; Al-Shehri & Stanley, 1992). This rapid expansion and massive increase in the quantity of medical education in Saudi Arabia has brought great challenges in relation to quality management and deserves to be shared with international scientific community, particularly in this era of globalization.

Two different opinions can be found in the literature about this expansion. One argues that Saudi Arabia needs reform similar to Abraham Flexner's recommendation (Atikan & Al-Mazroa, 2011), while the other affirms there is no need for such reform (Bin Abdulrahamn, 2011). Which argument should we accept? What are the underlying reasons for accepting one or the other? Are we talking about the same subject because the word "quality" may be perceived differently by different people in different cultures? Additionally, and more importantly, do we have a clear understanding of meanings and purposes of quality management in medical education?

2. Quality management in medical education: Meanings and purposes

Terms such as quality control, quality assurance, quality management and total quality management are commonly used in industry for the purpose of perfecting products (Gale & Grant, 2010; Hassan, 2010; Berwick, 1992; NHS, 2010). Quality control is a set of activities or techniques whose purpose is to ensure that all quality requirements are being met. In order to achieve this purpose, processes are monitored and performance problems are solved" (NHS, 2010). In medical schools, quality control means a curriculum implemented as planned, teaching formats and materials delivered according to standards, and learners monitored in terms of selection and compliance with the learning process. Quality control is about checking that the process meets predefined specifications. Quality assurance is the process of ensuring that the product or service is "fit for its purpose and not just made to the specifications" (Gale & Grant, 2010). In medical education, this means in addition to quality control there is a constant effort to improve and adjust for a better fit as the learning process goes on. Quality management includes quality control, quality assurance and all the activities that are carried out by all stakeholders at all levels and functions of medical education for the purpose of improvement (Ehlers, 2009; Gale & Grant, 2010; NHS, 2010). Quality management is a continuous and progressive cycle of improvement and further improvement.

However, most of the quality management processes in medical education are limited to quality control and accreditation which means achieving certain predefined objectives and criteria. This undoubtedly addresses one part of what quality management is about, but not entirely. Moreover, in developing countries, a culture of quality management is not well developed, hence, different stakeholders of medical education may perceive the purpose of quality management differently. Use of quality management in a mechanical way, from the 'green hill' of the manufacturing industry into the 'swamp' of medical education and health care, risks losing its main purposes of learning and professional development.

Ehlers calls for moving away from seeing quality management in education as a mechanical process that complies with pre-set standards, to a more holistic view that "focuses on change more than on control, development rather than assurance, and innovation more than compliance" (Ehlers, 2009). This latter approach stems from the concept that for quality management in medical education to be effective and useful, it has to emerge from within the educational organizations rather than be imposed from an outside source. This is in line with principles of adult learning and approaches of effective management and leaderships reported in literature (Brookfield, 1986; Kaufman, 2003; Al-Shehri & Khoja, 2009). Creating a culture of quality in medical education is an important requirement for stakeholders to realize the benefits and purpose of quality control, quality assurance and quality management. Stakeholders of medical education have to form a coalition of interest and build trust among themselves to enable them to address deficiencies and work as a team in a more transparent and productive way.

Unfortunately "medical education has failed to adequately prepare learners to function together effectively as members of an integrated health care team, with well-developed appreciation of the roles and responsibilities of every member of the team" (Cooke et al., 2010). This is compounded by lack of leadership styles and management structure that enable stakeholders in medical education and health care to realize their potential or face difficulties that affect their quality of work (Al-Shehri & Khoja, 2009). Accreditation is not usually designed to address different factors influencing quality improvement of medical education and accreditation cannot fully address different interactions among different stakeholders of medical education. Motivational, organizational and cultural factors play a significant role in the quality of medical education, but cannot be detected by predefined standards and guidelines (Ehlers, 2009). What can be checked on paper is not necessarily what is going on in reality. Stakeholders of medical education have to search continuously for excellence and further improvement.

Thus, the main purpose of quality management in medical education must be to promote a culture of excellence in which all stakeholders contribute safely, innovatively and competently to making the required change. This is particularly true in developing countries where conflicts among stakeholders are not uncommon in the absence of transformational leaderships (Al-Shehri, 2003; Al-Shehri & Khoja, 2009). The following four principles of quality management in health care mentioned by Berwick (1992) should be considered, bearing in mind the complexity and culture of medical education:

a. Meeting the needs of customers determines the success or failure of any organization or programmes. Responding to customers' needs is the quality against which organizations and programmes are judged. Needs change overtime and vary from culture to culture. Customers of medical education are not only students, residents and established doctors, but also patients, teachers, educators, managers and indeed the public. So which one of these customers determines the success or failure of medical education? Quality management in medical education is commonly defined by quality control and quality assurance through a process of accreditation according to the term of reference of the accrediting body. This process is not in line with principles of adult learning and may not necessarily lead to customer satisfaction.

b. Quality itself, regardless of its definition, is influenced by complex processes of production. In medical education these processes are related to physical, organizational and intellectual structures. The outcomes of these processes may be interpreted differently by different stakeholders (learners, teachers and faculty staff, and managers). Finding a method that addresses the needs of each stakeholder while ensuring quality is a challenge (Al-Shehri et al., 2008, 1994). But whatever method is used for quality management of medical education, it has to be meaningful and have a positive impact on people's health and a nation's costs. Accreditation cannot measure such impact. Involving all stakeholders in a more innovative and comprehensive way of developing a culture of quality management in medical education that enables them to examine issues of concerns critically and professionally with the purpose of improvement seems more appropriate than the mechanical approach of accreditation (Ehlers, 2009)

c. Professionals involved in the complex processes of quality management in industry, as in education, are intrinsically motivated to do well in perfecting the products. However, as indicated above in medical education the products (e.g. students in undergraduate medical education, trainees in postgraduate medical education, established doctors in continuing medical education) are also instrumental in perfecting the production process. Active participation of students, trainees and established doctors in the learning process not only affects outcomes, but is also an essential element of adult learning (Brookfield, 1986; Al-Shehri, 1992). Motivation of those professionals can be enhanced or reduced by beliefs, values, assumptions and behaviours that form the components of their culture (Schein, 2004; Miller et al., 1998). The culture of a group has been defined by Schein (2004) "as a pattern of shared basic assumptions that was learned by a group as it solved its problems of external adaptation and internal integration, that has worked well enough to be considered valid and, therefore, to be taught to new members as the correct way to perceive, think, and feel in relation to those problems" (Schein, 2004). If culture encourages trust among stakeholders of medical education to be inquisitive and transparent in their search for further improvement, quality will prevail. If on the other hand the culture fosters suspicion, blame and maintaining the status quo, then quality will diminish. Relying on the intrinsic motivation of professionals is not enough without creating the appropriate culture for such motivation to harvest its intention. In developing countries, unfortunately, culture is not usually conducive for talented professionals to sustain their intrinsic motivation to do well. This has to be changed by promoting a culture of quality management in medical education that is safe, innovative, inquisitive and systematic for professionals to sustain their intrinsic motivation in the search for perfection of their jobs.

d.   A strong built-in system of data gathering and analysis to monitor the process of production and link it to the outcomes is a continuous cycle known as the audit cycle. Such a system is essential to ensure improvement and meet the needs of all programmes. An audit cycle is usually used as a quality control measure by organizational management for process monitoring, but it can be used with other strategies to make quality management in medical education more purposeful for all stakeholders.

3. Strategy for creating a culture of quality management in medical education

Research in educational psychology has confirmed the effect of culture on education and mental activities: "how one conceives of education, we have finally come to recognize, is a function of how one conceives of the culture and its aims, professed or otherwise...culture shapes mind, that it provides us with the toolkit by which we construct not only our worlds but our very conceptions of ourselves and our powers...for you cannot understand mental activity unless you take into account the cultural setting and its resources, the very things that give mind its shape and scope" (Bruner, 1996). The importance of developing a culture of quality management that addresses the challenges facing medical education in different cultural settings cannot be underestimated by any means. However, a strategy of using effective means of learning and professional development may prove useful for stakeholders of medical education to promote a culture of quality management that is professionally led and scientifically sound. It can be called an ARR (Audit, Reflection and Research) strategy.

3.1 Audit

In addition to generating data for comparison with internal and external standards, an audit must be used as a strategy for self-directed learning and professional development. Defined as a learning and developmental strategy, the audit cycle shown in figure 1 can be applied to any particular experience of interest in medical education (Figure 1):

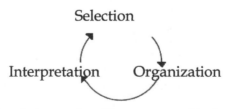

Fig. 1. 'Audit' as a learning strategy and quality management tool

Within the audit cycle, 'selection' is a key decision for the auditor. The whole curriculum or part of it can be selected for review; one or all of the teaching formats can be selected; part of the assessment or evaluation or the whole assessment methods and/or any other educational experience can be selected for auditing. The same can be applied to postgraduate medical education: selecting a particular experience relevant to the residency programme: rotation, tasks carried out by residents, governance and leadership of the programme, etc. In continuing medical education, selection may be related to one particular programme, all continuing medical education provision and/or impact on service/community, etc. Selection reflects the priority of the auditor.

However, before embarking on an audit project the following must be done: involving stakeholders from the beginning, checking resources available for the audit project and defining the potential benefits. Stakeholders are defined according to the relevance of the audited experience. This may include the dean of the college, chairman of the department under which experience is audited and/or a faculty follow who may not share your vision of the importance of the audit. Shared vision and involvement of stakeholders will optimize the benefits of the project (Al-Shehri et al., 1993). Checking resources is a straight forward reason to ensure that you have the manpower and materials necessary to conduct the audit project properly. The benefits of the audit need to be well communicated. This may be in the form of perceived shortcomings in existing performance and fears that these may have important consequences for the learning and teaching process or outcomes. An audit is a collaborative work that has to address an important experience relevant to different stakeholders and results in gains for all of them.

"Organization" may mean at one level the systematic collection of data on performance and on related standards, and at another level it may involve surveys and other measuring instruments, and a broader process of standard setting (Al-Shehri et al., 1993). This may include sorting out data from satisfaction surveys of learners (students, residents and established doctors), comparing findings with national and international standards and generating norm or criteria referencing for further audit projects. This process of gathering and sorting out data on the selected subject is in itself food for self-directed learning and professional development for all involved in the team, both individually and collectively.

"Interpretation" of results generated by the audit cycle deserves careful consideration. External standards such as those of the World Federation for Medical Education, the International Organization for Standardization (ISO) and other standards and guidelines (national or international) may be used as signposts for development rather than criteria that must be achieved. Deviation from such standards is not necessarily a reason for despair or joy, but rather a confidence indicator in the decision making process towards creating a culture of quality management in medical education. This may come in the form of collecting more data, introducing change and/or sharing the findings with others. The cycle of audit should be developmental where the findings of the audit lead to further development (Al-Shehri et al., 1993). An audit, then, should be a continuous journey rather than a project undertaken with the expectation of arriving at a predetermined station, which may not be the appropriate station for travellers. Stakeholders in different stages of medical education may undertake many journeys, some repeatedly to reach the required level and improve it further. Without a personal stake in the results of the audit, stakeholders will view it as an external quality control rather than a safe culture for more learning and further improvement. Perceiving quality management as a culture of learning and development should prevent resentment by some academics (Hassan, 2010) and motivate them to participate actively in producing a real culture of quality management.

3.2 Reflection

Reflection is another strategy which can be used to direct the purpose of quality management towards creating the right culture for stakeholders of medical education to be lifelong learners and critical thinkers in order to respond on time and appropriately to constant change in medical education and health care (Shon 1983, 1988; Al-Shehri, 1996). Reflection is defined in medical education as "a cycle of paying deliberate, systematic and analytical attention to one's own actions, feelings and thinking in relation to particular experiences for the purpose of enhancing perceptions of and responses to current and future experience" (Al-Shehri, 1996). Experience creates learning opportunities, but without a high level of awareness and readiness these opportunities may be lost (Kolb, 1984). Reflection is central to learning from experience and has been shown to enhance the competence of professionals. Developing new meanings from educational experiences through a cycle of selection, organization and interpretation (Figure 1) should enhance the quality of such experiences and take professionals to a higher level of cognitive and affective learning that enhance their competence (Al-Shehri, 1996).

Figure 2 shows levels of reflection where the reflective practitioner can develop from a lower level of learning and reflection (receiving and describing experience) into higher levels of synthesis and evaluation of experience. In this way professionals develop new meanings and perception that add to a better understanding and improvement.

Higher cognitive & affective learning skills

Fig. 2. Levels of reflection towards higher cognitive and affective learning

Systematic and structured reflection should enable reflective professionals to go from describing their experience to a higher level of synthesis and evaluation to add to the quality of current and future experiences. The following quotation from a reflective general practitioner shows how reflection can bring new meanings not only to the quality of service provided, but also the learning/training process behind that service: "I am often stuck by the effect on me of helplessness when I know that I cannot offer cure or treatment to a patient. We often state that patients want a simple answer to their problems - perhaps I may improve if I accept that there is more than just medical treatment involved in helping patients. Medical training does not equip one to accept this 'impotence'" (Al-Shehri, 1996).

Reflection and reflective practice has become a common term in medical education, but has not been utilized well for creating and developing a culture of quality management. There are many formats of reflection used in medical education that can help professionals to reflect systematically and structurally: random case analysis, group discussion, consultation analysis, using personal professional journals, portfolios, recording, talking and listening, and electronic reproduction. Each format has its pros and cons depending on the situation and resources. However, keeping a personal professional journal may prove simple and effective in enhancing professional development and competence (Al-Shehri, 1996). The presence of a mentor or a researcher may enhance further the development of reflection as a strategy for creating a culture of quality management in medical education. Unfortunately, little research has been done on the use of reflection in the quality management of medical education, particularly in developing countries, despite evidence of its role in self-directed learning and professional development. This takes us to research as the third strategy for developing a culture of quality in medical education.

3.3 Research

"Research in medical education is no longer in its infancy" (Norman, 2002). This is perhaps true in developed countries, but in some developing countries one could argue with confidence that research in medical education has yet to be born. Even in developed countries there is an acknowledgement that postgraduate training has not received the desired research (Schuwirth & Vleuten, 2006). In developing countries there is an urgent need to conduct research at all levels and in aspects of medical education to answer many questions in relation to quality management. The huge expansion and flourishing of medical education in terms of quantity in developing countries must be supported with research that addresses fundamental issues related to quality management. For research to be systematic and structural in promoting a culture of quality in medical education it could follow the selection, organization and interpretation process. Selection may address questions related to the curriculum management, learning and teaching process, and/or assessment in the undergraduate medical education. In postgraduate medical education, selection may address relationships between academic and health care providers, proper supervision and appropriate feedback to residents/trainees, opportunities given to master certain skills and procedures and best approaches to assessment and evaluation. For continuing medical education, selection may relate to finding evidence on cost effectiveness of continuing medical education provision, impact of continuing medical education on patients' care and service development.

The selection of a research project that responds to local needs and development would make stakeholders more interested in the organization and interpretation of data. Generating evidence-based data relevant to local needs, rather than to international or academic priorities, would guide the decision making process for quality management in medical education. Thus, interpretation of research findings should inspire stakeholders to create and maintain a culture of quality management in medical education rather than simply giving a certificate of success or a document of failure.

This shift towards creating a culture of quality management in higher education is new. "We are entering a new era in quality management for higher education. While it is difficult to mark its exact beginning, it is clear that it is moving away from a mechanistic to a holistic and cultural view of quality in education. It is characterized by an emerging understanding that quality development, in essence, demands for the development of an organizational culture based on shared values, necessary competencies and new professionalism" (Ehlers, 2009).

This is a welcome movement particularly for developing countries where culture determines to a great degree the meaning and perceptions of quality management. Three main strategies are described briefly above to make this movement more purposeful: audit the main engine for any quality assurance, reflection an innate but effective way of learning from experience and research to produce more meaningful data that illuminates both audit and reflection. The interaction of these three strategies can be exemplified by three circles continuously engaged with each other to identify common themes and priorities, but also keep each other moving for further improvement (Figure 3).

Fig. 3. Interaction of an ARR strategy

The process of selection, organization and interpretation should help in making these strategies more systematic and structural for the purpose of developing a culture of quality management in medical education that takes into consideration internal and external factors. External standards (national or international) are certainly useful, but need to be used as signposts for educational and professional development, not as edicts to be shoved down the throats of stakeholders. Once the culture of quality management has become an integral part of the continuum of medical education and familiar and acceptable to all stakeholders, then external accreditation ceases to be a burden, but rather an enjoyable exercise.

4. Quality management and medical education in Saudi Arabia

In 2004 the National Commission for Assessment and Academic Accreditation (NCAAA), the national body responsible for quality management in higher education, was established as a result of a Higher Education Council resolution under the supervision of the Ministry of Higher Education. In 2006 the Central Board of Accreditation of Healthcare Institutions (CBAHI) was formed under the Ministry of Health with the aim of setting standards and accrediting hospitals and health care centres (www.cbahi.org.sa). In general therefore the concept of quality management in higher education and health care is relatively new to Saudi Arabia (SA) in terms of structure and organization. This does not mean medical education and health care specialties in SA were not aware of the concept of quality management before the establishment of the NCAAA and the CBAHI as the Saudi Council for Health Specialties (SCHS) was established earlier (1992) and responsible for the accreditation of postgraduate medical education and continuing medical education (Al-Shehri et al., 2001; Al-Shehri et al, 2008).

Moreover, many of the medical service organizations (e.g. hospitals and specialized medical centres) were based on models taken from developed countries that recognize quality management and many of the health care professionals working in these medical organizations were trained in well-recognized training centres abroad that embrace quality management. So as a concept, quality management has been known about in medical education and health care in Saudi Arabia for decades. Indeed, some well-known specialized hospitals in Saudi Arabia were accredited by the American Joint Commission and other international health care accrediting bodies, and some of the old medical schools had been accredited at one point of their development by an international educational accrediting body. However, as a national body and structure, quality management in higher education and health care is a recent movement in SA marked by the establishment of the NCAAA and the CBAHI, respectively.

This is a welcome movement, but it seems that the number one priority for the NCAAA is accreditation rather than quality management (Tekian and Al-Mazroa, 2010). The NCAAA "has been established...with responsibility to establish standards and accredit institutions and programs in post secondary education. The system for quality assurance and accreditation is designed to support continuing quality improvement" (NCAAA, 2009, http://www.ncaaa.org.sa. Accessed 4 Sep 2011). The assumption is that accreditation will lead to quality improvement. The CBAHI has accredited more than 21 hospitals under MoH so far and 60 more hospitals are under consideration (Al-Jazirah Newspaper, 2011) but little evidence can yet be found on the impact of such accreditation on customers' satisfaction as almost daily there are articles in newspapers reflecting customers' dissatisfaction with services in the accredited hospitals. The same can be said about the accreditation process of continuing medical education activities by the SCHS: it does not address the real issues of quality management (Al-Shehri et al., 2008). As indicated earlier, setting standards and accreditation is a quality control issue that does not necessarily lead to quality management of medical education.

It is understandable that quality control is a priority for the NCAAA, the CBAHI and the SCHS, but it is important at this stage to learn from the shift towards a holistic approach to quality management in medical education (Ehlers, 2008). Accreditation and standards setting is only one part, albeit an important part, of quality management. So far the NCAAA has done a good job by conducting workshops, courses and orientation sessions for faculty staff and stakeholders of higher education to educate and orient them about the accreditation process, criteria and standards. This should be encouraged and developed further towards establishing a culture of quality management rather than how to be accredited. Unless there is a clear distinction between quality control and quality management there is a danger of seeing quality as a destination (getting accredited) rather than a journey (culture of continuous improvement and innovations). Medical education and health care are evolving quickly and vary from one place to another. Areas of concern recognized a decade ago may not necessarily be relevant now.

4.1 Undergraduate medical education

In 1999 the following issues related to undergraduate medical education in SA were reported (Al-Shehri & Al-Ghamdi, 1999):

- Overload of curriculum with information and factual knowledge that have little relevance to actual challenges and trends in real clinical practice.
- Lack of important subject areas like communication skills, medical ethics, managed care and leadership, psycho-social factors affecting people's health and use of bio-psycho-social models in problem solving and problem-based and community-oriented learning.
- Lack of proper support for secondary school graduates to cope with their new professional learning in medical schools dealing with the most sensitive issues with people's health and well-being.
- Lack of consensus on the core curriculum and subjects of interest for students.
- Passive teaching where teachers transmit information through lecture with little, if any, active participation from students.
- Teacher-centred and rote learning approach is predominant. Students memorize teachers' talks by heart and reproduce them in examinations.
- Producing dependent rather than self-directed learners.
- Classrooms and hospital wards are the context in which most, if not all, undergraduate teaching takes place. Students and their teachers may lose touch with the reality of the outside world. It is well-known that neither patients seen in teaching hospitals are representatives of their community, nor is their presentation typical of natural display.

In recent publications there is evidence of change. Medical schools in SA are acknowledging the recent trends of problem-based and student-centred learning (Bin Abdulrahamn 2008, 2010) and courses on medical ethics and communication skills are provided (Haqwi & Al-Shehri, 2010). However, there is little research-based evidence to support the quality of this apparent change. Moreover, new challenges emerge as a result of the speed with which new medical schools have been established; these include not enough qualified faculty staff, lack of evidence on the best curriculum to adopt, shortage of teaching and training facilities, huge intakes of medical students with unclear evidence of selection and the absence of a clear management plan for faculty staff development in order to deal with issues arising from the teaching-learning process and outcome (Tekian & Al-Mazroa, 2011). Also, Azer (2007) raised the following questions in relation to undergraduate medical education. "What evidence do we have that PBL curricula foster self-directed learning? What will make a good medical curriculum? What actions should we take to ensure better learning in the PBL structure? What is the use of good role models in teaching? What tools could we use in the assessment of non-cognitive competencies such as interpersonal skills, empathy, integrity and effective communication?"

To use industry terms, lines of production in undergraduate medical education are: "selection, mentoring, tutoring, teaching, learning, participation, assessment and examinations." Lack of quality at any station or step will result in a substandard product (Gale & Grant, 2010). The current system of selecting medical students depends mainly on secondary school grades, national aptitudes and achievement examinations (Al-Alwan, 2009). More research-based evidence is needed to support this approach. "The leap of logic that equates high marks in an examination at the terminal end of adolescence with a human and caring medical profession is nonsense" (Best, 1989). Moreover, lines of production cannot be addressed in the absence of a holistic view of the culture in which medical education is taking place.

Accreditation using internal or external standards in a summative way will be able to identify problems at each point in lines of production, but may result in drastic consequences that do not match the society needs of expansion in medical schools. For example, Flexner's reform resulted in reducing the number of medical schools in the USA at that time from 155 to 31; is this what we want in SA? Even if this drastic step is accepted, what evidence do we have to equate accreditation with quality management? There must be a way of introducing quality management that maintains recent needed quantity development of undergraduate medical education in SA while ensuring that the product of medical schools meets customers' needs.

Using "Audit", "Reflection" and "Research" (ARR strategy) systematically and structurally for the purpose of developing a culture of quality management in medical schools where all stakeholders participate actively and meaningfully seems more appropriate at this stage. Accreditation should be used to enhance such development rather than distorting or stagnating it. The purpose of quality management in undergraduate medical education should be to show evidence of development in the right direction. Stakeholders must be encouraged to participate actively in auditing, reflecting and researching their issues of concern to show that today is better than yesterday and tomorrow will be better than today, rather than getting a certificate that has an expiry date. The NCAAA may rethink the way it plans accreditation to make it more informative to the development of a culture of quality management in undergraduate medical education, rather than a summative assessment of medical schools.

4.2 Postgraduate medical training

Until 1990 most of postgraduate medical training was conducted abroad: graduates were sent to well-known training centres in developed countries. This resulted in many doctors who had graduated from different and well-known medical centres all over the world coming back to SA to become leaders in their field of expertise. This diversity is an advantage that should be used and built upon. However, it is important also to remember that mastering one field of medical specialty does not mean mastering the training process. Trainers need to be trained further, particularly in relation to medical education. Now most postgraduate training take place inside SA with more than 21 Board training programmes and 39 Fellowships run annually at different training and academic organizations under the supervision of the Saudi Council for Health Specialties. The organizational structure, management, professional supervision and monitoring, evaluation and assessment, and communications among stakeholders of these programmes deserve to be audited, reflected upon and researched for the purpose of creating a culture of quality management in postgraduate medical education for the purpose of improvement.

At the Saudi Council for Health Sciences, each programme is represented by a Board. This Board is responsible for setting standards and accrediting relevant programmes under its specialty. For example, the Saudi Board of Family Medicine is responsible for approving any residency programme in family medicine according to the Board's standards. No programme will start without the approval of the relevant Board on the basis of accreditation procedures for the training centre and the hospital. However, once a programme gets accredited and is given the permission to start training, there is no strong accountability and responsibility structure to ensure the presence of quality management in the residency programme at different accredited training organizations. It is left to the accredited programme to run the show. This may be good in encouraging individuality and innovations, but again it indicates the shortcoming of accreditation as a ticket of entry, but no guarantee of the quality of process and outcome.

Few publications can be found on quality management of postgraduate medical education in SA, but common observation suggests that it deserves better attention. In fact the only publication found suggests that the satisfaction of residents/trainees with one of these programmes is very low (Alghamdi, 2008). This supports the argument that quality management is more than setting and complying with standards; it is a culture that must be developed on many fronts. Using the ARR strategy should tell us more about the current experience and how to go forward if used systematically and structurally to address issues of concerns. Anonymous residents' surveys, evidence on proficiency of trainers and plans taken to enhance their proficiency further, mortality and morbidity reports of training centres and hospitals, reports on governance and leadership, exposure and experience of residents at different levels with appropriate supervision and mentoring, patients satisfaction surveys and many other evidence and data should be generated and utilized to enhance the culture of quality in postgraduate training programmes.

The ARR strategy can be used to look into the above mentioned and other issues professionally and scientifically to generate sound evidence to help in quality management of postgraduate medical education. Another important issue that needs to be looked at is the gap between undergraduate medical education and postgraduate medical education. Internship is a clear example of this gap that must be bridged.

4.2.1 Internship

Graduates of medical schools have to spend one year (internship) in a training hospital before they join any postgraduate training programmes. This year is a must for graduates in order to be awarded their Bachelor degrees. This internship year is a critical period of joy and uncertainty. A joy for graduates celebrating their final step after a long journey of at least six years spent in undergraduate medical education and uncertainty because the majority of graduates are not sure of their next step towards specialization. It is a common observation that in our system this period spent in two extremes. On one extreme it is a year of hard work in which intern suffers from routine and tough rotations, on the other extreme it is a year of relaxed 'fly on the wall' observation without involvement and supervision. This is mainly due to the gap in management and communication between academics in medical schools and health care providers in the training hospitals used for internship. Teaching hospitals under medical schools cannot cope with the expansion of medical schools and high number of graduates. Thus, service hospitals are used for internship rotations, but medical schools have little say on the management and supervision of interns in these service hospitals. Interns need proper supervision and monitoring to ensure that the internship period consolidates what they have learned in medical schools and prepares them for further training.

During internship new graduates of medicine may need support in dealing with their emotions and uncertainty in coping with new experiences and future aspirations for their careers. If left unattended by experienced academic supervisors and mentors, interns may lose interest or acquire bad learning habits. If overworked and stressed they may not realize the joy and potentials of being graduates of medical schools and future doctors. Balancing the two extreme is very crucial to make the internship period a useful base for residency programmes and inspiring intern's confidence and responsibility. One way of helping interns with the transition from being students to residents with more responsibilities is half day release activities in the presence of experienced academic facilitators where the interns come together out of clinical rotations to discuss issues that stem from their experiences, emotions and feelings, worries and uncertainties. Close supervision and management of interns would ensure a balance between the two extremes mentioned above and may detect the need for counselling sessions with certain interns going through stressful situations. The death of a patient, a medical error, conflict with team members and other stressful situations may leave long lasting effects on interns without noticing. The half day release may help in discussing these issues, but counselling services may also be needed.

Another issue of concern with internship is 'the fixed truth' about the need for interns to have rotations in the four major specialties: medicine, surgery, paediatrics and obstetrics, and gynaecology. I do not know on what basis these rotations become fixed truth. There are graduates who have already made their minds up about a residency programme outside these specialties. We need evidence to support the benefits of these traditional rotations. Besides, common sense and experience tell us that family medicine and primary care may be more appropriate for interns in order to come to understand the most common problems that exist in the community and how to deal with them (Al-Shehri, 1997; Al-Ghamdi & Al-Shehri, 1996). I would argue that a rotation in a well supported primary care centre is an essential experience for all interns.

Again we do not have data on which we can make the right decisions. The ARR strategy should look critically into the internship period in terms of structure, organization and management, and outcomes to generate strong evidence on quality management of this essential period. Lack of enough teaching hospitals dedicated to colleges of medicine and the gap in organization and management between undergraduate medical education and postgraduate medical education may compromise the quality and outcome of the internship. Primary care centres and family medicine clinics have not been well utilized for interns. The process of communication and supervision between colleges and hospitals are not well established and depend on informal collegial relationship more than anything else. The expansion of colleges and the increase in the number of graduates is far greater than current teaching hospitals can accommodate.

The challenge is more for private colleges as they cannot access any of the public teaching hospitals and have not established their own hospitals. Private colleges may make use of the flexibility of being in the private sector and make innovative moves toward community-based teaching and training. They may own or manage family medicine and primary care centres, and use them as teaching sites in addition to teaching hospitals. The private sector should be less bureaucratic than the public sector in making and managing change. Pursuing community-based teaching and training of medical students and interns by the private sector may open up new territory for the public sector to follow.

4.3 Continuing medical education

In Riyadh, the capital city of SA, it has been estimated that more than 8,000 continuing medical education programmes are provided annually which are accredited centrally by the Saudi Council for Health Specialties . The quality management of this huge continuing medical education provision is described as "at best unclear with little impact on stakeholders, healthcare services and patient care, and at worst encouraging "cheap junk" continuing medical education" (Al-Shehri et al., 2008). After reviewing educational models of evaluation that can be used or modified in the quality management of continuing medical education provision, the authors emphasized "that one cannot decide on a type of evaluation model without understanding the stakeholders of continuing medical education as adult professionals and without understanding the complexity of the assessment process in terms of validity, reliability, feasibility and acceptability. The difficulty of finding valid, reliable and feasible models of evaluation that takes into consideration the interests of all stakeholders as adult learners and at the same time addresses the knowing, feeling and doing of stakeholders cannot be underestimated" (Al-Shehri et al., 2008).

The ARR strategy can help in looking into different issues of concern regarding quality management of continuing medical education, including innovative evaluation models of continuing medical education that enhance learning and professional development of doctors, while benefiting patients care and health care systems. Research centres must fund large and innovative projects on quality management of continuing medical education. A combination of research findings with auditing projects and reflective practice must lead to a culture of quality management in continuing medical education that ensures its merits and outcomes.

5. Conclusion: Closing the loop

More than 100 years ago Flexner's report changed medical education in the USA and consequently worldwide. Against a predetermined gold standard all medical schools were judged and this judgment resulted in closing more than 50% of the medical schools. "Flexner had determined that the template for judging all medical schools would be Johns Hopkins with its academic rigor, its teaching hospitals and the quality of its full-time faculty" (Cook et al., 2010). However, recently, the same institution behind Flexner's report

(The Carnegie Foundation for the Advancement of Teaching, USA) called for reforming medical education in the USA, but in a different way: "those who teach medical students and residents must chose whether to continue in the direction established more than a hundred years ago or take a fundamentally different course, guided by contemporary innovation and new understanding about how people learn...new discoveries in the learning sciences and changes in the preparation of physicians all argue for the need to reexamine medical education" (Cook et al., 2010).

The lessons for developing countries, particularly SA, is that predetermined standards and accreditation will definitely contribute to quality management of continuing medical education, but leave much to be desired. In fact accreditation in a culture where quality management is not well understood may do more harm than good (Ehlers, 2009; Schein, 2004; Hasan 2010; Stanley & Al-Shehri, 1993). A culture of quality management in medical education has to be developed first in which all stakeholders contribute and innovate safely. In this way accreditation and meeting external standards ceases to be a threat, but rather becomes an opportunity for further learning and development.

The ARR strategy is proposed as a powerful toolkit to develop a culture of quality management in medical education in SA. This strategy has shown to enhance self-directed learning and professional development among professionals, and has the potential of addressing issues of quality in the continuum of medical education over time. Stakeholders of medical education in SA should work together in a team at all levels and functions to generate data, review them and implement required change in a developmental process. Prioritization of issues and needs relevant to different stakeholders can be addressed by following a systematic and structured way through the process of selection, organization and interpretation. In this way the development of medical education in SA can be guided safely in searching for the highest quality without jeopardizing what has been achieved in quantity. National standards and criteria (e.g. those used by the NCAAA, the CBAHI and the SCHS) as well as international standards and accreditation (e.g. WFME, ISO and others) should be used as signposts leading to the right direction rather than secured gates through which permission to travel is granted.

Using audit, reflection and research as shown above (Figures 1, 2 and 3) should engage the hearts and minds of all stakeholders of medical education in SA to develop a culture of quality management that transforms current experiences into products that have a positive impact on the nation's health and economy.

6. References

Al-Alwan I, (2009). Association between scores in high school, aptitude and achievement exams and early performance in health science college. *Saudi Journal of Kidney Diseases and Transplantation, 20(3), 448-453.*

Al-Ghamdi KM, (2008). Current status of dermatology residency training in Saudi Arabia: trainees' perspectives. *East Mediterr Health J*; 14(5): 1185-91.

Al-Haqwi A & Al-Shehri, (2010). Medical Students evaluation of their exposure to the teaching of ethics. *Journal of Family and Community Medicine*; 17: 41-45

Al-Shehri AM & Al-Faris E, (1998). Learning objectives of medical students: what is their message? *Saudi Medical Journal; 19(1): 747-749.*

Al-Shehri AM & Al-Ghamdi A, (1999). Is there anything wrong with undergraduate medical education? (leading article) *Saudi Medical Journal; 20(3)215-218.*

Al-Shehri AM & Stanley I, (1993). Continuing Medical Education for Primary Care: problems and solutions. *Saudi Medical Journal; 14:8-11.*

Al-Shehri AM & Stanley I, (1993). Developing organizational vision in General Practice. 2. *British Medical Journal; 307: 101-103*

Al-Shehri AM, (1992). General practitioners' PGEA: The first six months (survey in Mersey Region). *Postgraduate Education for General Practice; 3:53-61.*

Al-Shehri AM, (1992). The market and educational principles in continuing medical Education for general practice. *Medical Education; 26:384-388.*

Al-Shehri AM, (1995). A model of self-directed learning for established practitioners. *The Bulletin of National Guard Health Affairs; 6(49): 13-14.*

Al-Shehri AM, (1996). Learning by reflection in general practice: a study report. *Education for General Practice; 7:237-248.*

Al-Shehri AM, (1997). Primary Care-led NHS: what does it mean? *Journal of Kuwait Medical Association; 29:265-268.*

Al-Shehri AM, (2003). Is there a need to involve doctors in the management of healthcare in Saudi Arabia? *SMJ; 24(11): 1165-1167.*

Al-Shehri AM, (2010). E-learning in Saudi Arabia: "To E or not to E", that is the question. *Journal of Family & Community Medicine; 17(3):147-150*

Al-Shehri AM, Al-Haqwi A, Al-Gamdi A & Al-Turki S, (2001) (leading article). Challenges facing CME and Saudi Council for Health Specialty. *Saudi Medical Journal; 22 (1):3-5*

Al-Shehri AM, Bligh J & Stanley I, (1993). A draft Charter for general practice continuing education. *Postgraduate Education for General Practice; 4: 161-167.*

Al-Shehri AM, Bligh J & Stanley I, (1994). Evaluating the outcomes of continuing education for general practice: a coalition interest. *Postgraduate Education for General Practice; 5: 135-142.*

Al-Shehri AM, Stanley I & Thomas P, (1993). Continuing education for general practice. 2. Systematic learning from experience. *British Journal of General Practice; 43: 249-253.*

Al-Shehri, AM & Khoja T, (2009). Doctors and Leadership of Healthcare Organizations. *Saudi Medical Journal; 30 (10): 1253-1255*

Al-Shehri, AM, Al-Haqwi AI & Al-Sultan M, (2008). Quality Issues in Continuing Medical Education in Saudi Council Arabia. *Annals of Saudi Medicine; 28(5) 378 – 381.*

Azer SA, (2007). Medical Education at the Crossroads: Which Way Forward. *Saudi Medical J; 26(3):153-157)*

Berwick DM, Enthoven A & Bunker JP, (1992). Quality management in the NHS: the doctor's role – I. *BMJ 304: 235-239.*

Best J, (1989). The politics of the sand-pit. *Med J Aust 150: 158-161.*

Bin Abdulrahman K, (2008). The current status of medical education in the Gulf Cooperation Council countries. *Ann Saudi Med; 28(2):83-88.*

Bin Abdulrahman K, (2011). Saudi Arabia does not need an Abraham Flexner. *Medical Teacher; 33: 74-75.*

Brookfield S, (1986). *Understanding and facilitating adult learning.* Milton Keynes: Open University Press.

Bruner J, (1996). *The Culture of Education.* Harvard University Press, USA.

Cooke M, Irby D & O' Brien BC, (2010). *Educating Physicians: A Call for Reform of Medical School and Residency.* The Jossey-Bass business & management series. USA

Cueto, Burch & Adnan, (2006). Accreditation of Undergraduate Medical Training Programs: Practices in Nine Developing Countries as Compared with the United States etal *Education for Health; 19(2): 207-222*

ECA, *(2005).*

Ehlers UD, (2009). Understanding quality culture. *Quality Assurance in Education; 17: 343-363*

Fraser R, (1991). Undergraduate medical education: present state and future needs. *British Medical Journal*; 303: 41-43.

Gale R & Grant J, (2010). Quality Assurance systems for medical education. 13: 113-120.

Hassan T, (2010). Doctors or technicians: assessing quality of medical education. *Advances in Medical Education and Practice; 25-29.*

Kaufman DM, (2003). Applying educational theory in practice. *BMJ; 326: 213 - 216*

Kolb D, (1984). *Experiential learning.* Eaglewood Cliffs, NJ: Prentice-Hall.

Madison D, (1978). What's wrong with medical education? *Med Educ; 12:97-102.*

Miller J, Bligh J, Stanley I & Al-Shehri AM, (1998). Motivation and continuation of professional development. *British Journal of General Practice; 48: 1429-1432.*

Modernizing Medical Careers. MMC Inquiry.http://www.mmcinquiry.org.uk/draft.htm. accessed 26 Sept 2011.

Moust JHC, Van Berkel HLM & Schmidt HG, (2005). Signs of erosion: reflections on three decades of problem-based learning at Maastricht University. *Higher Education; 50; 665-683.*

NHS, (2010). Quality Management of Postgraduate Medical Education and Training in Scotland. www.nes.scot.nhs.uk/medicine

Norman G, (2002). Research in medical education: three decades of progress. *BMJ 324 (7353); 1560-1562.*

Powis DA, (2008). Selecting medical students. *MJA; 188: 323-324.*

Schein EH, (2004). *Organizational culture and leadership.* Third Edition. The Jossey-Bass business & management series. USA

Schon D, (1983). *The reflective practitioner: how professionals think in action.* New York, NY: Basic Books.

Schuwirth & Van der Vleuten C, (2006). Challenges for educationalists. *BMJ; 333 (7567): 544-546*

Sood R & Adkoli BV, (2000). Medical Education in India – Problems and Prospects. *Journal, Indian Academy of Clinical Medicine. 1(3):210-213.*

Stanley I & Al-Shehri AM, (1992). What do medical students seek to learn from General Practice? *British Journal of General Practice; 42: 512-516.*

Stanley I & Al-Shehri AM, (1993). Reaccreditation; The why? What? And how? Questions. *British Journal of General Practice; 43:524-529.*

Stanley I, (1993). Continuing education for general practice. 1. Experience, competence and the media of self-directed learning for established general practitioners. *Br J Gen Pract; 43:210-214.*

Tekian A & Al-Mazroa AA, (2011). Does Saudi Arabia need an Abraham Flexner? *Medical Teacher; 33: 72-73.*

Tooke J, (2008). Aspiring Excellence. Final report of the independent inquiry into
Woodhouse D, (2004). The quality of quality assurance agencies, *Quality in Higher Education,*
       10(2): 77-87.

7

Organizational Service Orientation as a Quality Predicator in Services

Wieslaw Urban

Bialystok University of Technology, Faculty of Management
Poland

1. Introduction

The concept of the "service economy" was reviled many years ago, but some crucial issues like service quality are still interesting as research fields. Now, in the globalized, web 2.0 economy, services are an area of many outstanding challenges to managers. The old question very often asked by managers, "How can I delight my customers?" sounds every time fresh and curiously. Even in the digitalised world, when customer service staff use intensively electronic devices while serving customers, quality is still crucial and challengeable, because on the two sides of a Skype connection or an e-mail exchange they are real humans. Staff can provide outstanding service via the latest Facebook communication interface and delight their customers, or staff can provide poor service via the same channel disregarding customers, and treating them as petitioners. Customer services are taking on an ever-increasing level of importance in today's global economy (Baydoun et al., 2001). Furthermore, we can observe the growing popularity of the service approach to management. According to Vargo & Lusch (2004), service logic is valid not only in what have traditionally been called services, but also in all industries in the economy.

A service product must be provided well first time, and there is no room for failure because service production takes place at the interactive with customers. The quality of interaction in the service encounter, and customer service experiences, lead to customer loyalty, and are very often treated as the key factors of a successful business. A service organization must be fully prepared to provide excellent service. On the other hand, managers need to know the capability of an organization to provide superior services. In this context there is the very important task of searching for organizational predicators of excellent outcomes provided to customers, as well as assessment methods of an organization's ability to provide excellent services. That is why the organizational service orientation concept is proposed for deep examination in this chapter.

In the stream of different concepts which are trying to assess an organization's ability to provide excellent service outcome the idea of organizational service orientation seems to be particularly useful, but it is not a widely practiced approach among managers and researchers. Organizational service orientation is very closely related to the corporate culture concept, and describes a staff approach which is directly connected to quality of service delivery; it determines the state of all interactions between an organization and its customers.

This chapter elaborates on the problem of organizational service orientation and a few related concepts. There is an attempt to evaluate the state of organizational service orientation in the service industry in Poland. The cross-sector approach is employed, and the problem of the differences of organizational service orientation in different service sectors is examined. A wide cross-sector study has not appeared before in an investigation of organizational service orientation. The chapter provides an exhaustive analysis of some theoretical concepts related to organizational service orientation, which may be valuable for a better understanding of the assumed research approach.

2. The link between inside and outside service quality

The classic approach to quality management, namely TQM, is mostly focused on quality in the internal meaning. The excellence of activities done by people from inside a company this is the main aim in this approach. Nevertheless, in service research it is noticeable that most studies have considered only quality experienced by customers (Mukherjee & Malhotra, 2006). Very often the discussions about quality in services exclusively take into consideration the customers' perception context. The meaning of quality as customers' experiences is different from the organizational one because this is an individual and subjective concept. So, in the literature output we have a few models trying to link inside service quality with the quality experienced by customers, i.e. outside service quality. Researchers are also looking for the organizational predicators of quality experienced by customers, as well as the relationships between the state of the organizational service system and the effects on business.

The widest known concept which links the internal and external sides in services is given by Heskett et al. (2008). This is the Service-Profit Chain concept, where company profit and precedent customer satisfaction are affected by the service delivery system, and all have their sources in internal service quality. The concept underlines the fundamental meaning of the inside quality of activities in an organization while building customer satisfaction and, finally, profitability. Another theory explains customer perceived quality by gaps which appear in a service organization. This is the five quality gaps model proposed by Parasuraman et al. (1988). According to the authors the quality experienced by customers depends upon the information loop passing through an organization, which informs us about the customers' real expectations. Four gaps influence the customers' perception of quality. The first one exists in managers' perceptions about what customers desire from services. The second one is contained in the specifications prepared by managers. The third one is how the specifications are delivered to customers, and the final one appears in external communications giving information on the service quality that a company offers to deliver.

There are many more pieces of research which look for predicators of customer service quality and customer satisfaction in many different spheres of an organization. According to some authors, customer-contact employees' job standardisation might lead to a higher level of service quality experienced by customers (Hsieh & Hsieh, 2001). Others underline the role of staff empowerment in achieving high service quality (Ueno, 2008). Management techniques (Kantsperger & Kunz, 2005) and employees' effort, involvement and abilities (Specht et al., 2007) are also treated as factors predicating customers' perceived quality. Johnston & Clark (2005) underline the importance of the service processes in creating the customers' service experience. Also Urban (2011) proposes the service quality measurement model which simultaneously evaluates organizational service processes and customers' perception.

Another bundle of factors leading to customers' superior experiences lies in the corporate culture of a service organization. Schneider et al. (1998), grasping what is specific in corporate culture to a service organization, exhibit how the service climate leads to customer perceptions of service quality. The issue of searching for predicators of customers' superior experiences in close-to-culture spheres of the service organization is the leitmotif of this chapter.

3. Organizational culture in services

Among the many factors affecting market and financial success of an organization, the culture code within an organization is considered an important predicator of market and financial performance. In the classic work "In Search of Excellence" by Peters & Waterman (1982) the authors conclude that the most successful companies have expressive and clearly defined organizational cultures. The authors clearly indicate that any organization in order to survive and achieve success must have a specific set of beliefs and values. These beliefs and values influence the policies and actions of companies. These beliefs and values are very stable. Almost anything can be changed while adjusting an organization to forthcoming challenges, but they are constant in the long term.

Hofstede (1997) sees values as the centre of the culture. From values emerge practices which are symbols, heroes and rituals. Values are closely linked to a tendency for certain decisions to be made by people. According to this author the organizational culture differs mainly at the level of practices which are characteristic for particular organizations. Others, for example, propose that organizational culture refers to values in the context of expectations, support and rewards in an organization, as well as norms on which polices, practices and procedures in an organization are based, and finally the common, shared interpretations of values and norms within an organization (Schneider 1988). Values and norms are not directly observable in an organization, but we can see practices, procedures and behaviours driven by them. In the culture phenomenon visible and invisible elements coexist in mutual interdependence. Culture is also the essence of organizational identity. It provides the basic guidelines for employees, and that is why it is called the collective programming of the mind.

In the literature output we can find the specific interpretation of organizational culture in particular service industries. Beitelspacher et al. (2011) suggest that service culture in a retail organization should be meant as a customer-centric culture, which aims at exceeding customer expectations, and is strongly focused on providing superior customer value. If the culture is beneficial, it should be strongly associated with the knowledge and skills development of employees.

The positive influence of service culture on the key performance variables of an organization can be observed. Empirical research points out that first of all it leads to better customer perception of service quality, as well as having a beneficial impact on customer market loyalty, repurchase intentions and customer satisfaction (Beitelspacher et al., 2011). This research, conducted in the retail industry, underlines the importance of the issue of organizational culture in the service context. Culture programming appears to be a meaningful issue for the endurance and long term development of a service enterprise. The positive relationship between organizational culture in a service organization and its performance, namely output quality and customer satisfaction, has been demonstrated by several studies (Winsted, 1997; Furrer et al., 2000; Kanousi, 2005).

Researchers also underline some differences between the corporate culture in the general meaning and the service sense of corporate culture, so called service-oriented culture. Service-oriented culture has particular significance in manufacturing. According to Gebauer et al. (2010) it consists of service-oriented values and behaviours; the authors distinguish the values and behaviours of management and employees. Service-oriented aspects in the background of corporate culture refer to entrepreneurial orientation, real problem-solving eagerness, innovativeness, and employee flexibility. According to studies performed by authors, service-oriented elements in organizational culture are key factors in achieving sustainable business performance.

As we see from this short glance at organizational culture phenomenon, it includes significant factors which influence the organization's output, and it plays an important role in the long-term business success of an enterprise. It also offers a broad framework for particular conceptualization of this phenomenon for particular purposes. In this study the attention is focused most of all on the organizational service orientation concept.

4. Organizational service orientation

Among culture-originated concepts, which express an organization's ability to provide excellent service to customers, organizational service orientation concept seems to be very accurate and relevant. Organizational service orientation manifests itself in staff attitudes and behaviours which directly affect the quality of the service delivery process, and determine the state of all interactions between a service organization and its customers. An organizational service orientation is defined by Lytle et al. (1998, p. 459) as an organization-wide embracing of a basic set of relatively enduring organizational policies, practices and procedures intended to support and reward service-giving behaviours that create and deliver service excellence. At the visible level it is reflected by genuine attention to customer needs, as well as sharing, helping, assisting, and giving support to customers. Organizational service orientation is recognized as the kind of predisposition for giving superior service. Its supposed direct impact on the state of service provision makes this concept very interesting and potentially valuable.

The service orientation stays in a strong relationship with the intangible aspects of an organization. It exists when the organizational climate for service crafts, nurtures and rewards service practices and behaviours known to meet customer needs (Lynn et al., 2000, p. 282). It is also taken as something that manifests itself in the attitudes, as well as actions, of members of an organization which values highly the creation and delivery of excellent service (Yoon et al., 2007, p. 374).

According to Lytle et al. (1998) an organizational service orientation consists of fundamental elements which cover four delivery fields crucial for service, which are: (1) service leadership practices, (2) service encounter practices, (3) service system practices, and (4) human resource management practices. The proposed fields are fulfilled by specified elements which constitute the best-in-class service practices and procedures. So that the organizational service orientation is a particular pattern of the best, as far as possible, approach to giving superior value to customers.

Leadership is treated by many management theories as the first necessary condition for sustainable organization growth. Along with leadership, very often the strong and long-reaching vision of an organization is mentioned as a critical success factor. Lytle et al. (1998) mention the particular importance of servant-leaders in the organizational service orientation. The direct engagement of servant-leaders in helping and assisting personnel leads to superior service; it builds special kind of unwritten standards informing staff how to perform a service. The service vision, which might be perceived as a kind of service manifesto, informs the whole staff on long-term objectives and goals.

The service encounter field refers to customer treatment and staff empowerment. How a service provider looks after customers is the first and the most important predicator of the quality perceived by them in many service industries. In the literature output there is a conformity of opinion that says that to get delighted customers it is required to allow direct contact staff to act with very unconstrained manners. Only in this case will employees be able to react flexibly to customers' needs and provide superior service.

All the service provided to customers exists in, and is produced by, a system. According to Deming, the most renowned quality management guru, in a system we should be looking for the potential for quality improvement as well as waste elimination (Deming, 1994). The service system has peculiar components that play a fundamental role in achieving service quality. One of them is failure prevention, and the second is service recovery. Traditionally failure analysis is distancing between internal and external failures; in service we have almost exclusively external failures – all the service production is carried out with the participation of clients. This clearly increases the importance of failure avoidance. On the other hand a failure might be an excellent chance to delight a customer, provided that a company has an effective, workable system to deal with complaints and faults. Service recovery might be a chance to win customer loyalty, including word-of-mouth (Swanson & Kelley, 2001). According to Lytle et al. (1998) there are two more crucial factors from the point of view of achieving a high service quality system: technology utilisation and dissemination of service standards throughout an organization.

Most services are work intensive; the responsibility for the treatment of customers lies primarily in staff's hands. Individual personal skills, professional preparation, type of personality, mood and many other personal factors affect customers' experiences of quality. The employees' training and appropriate motivation process seem to be very beneficial in the field of human resource management practices.

Researchers have been asking themselves whether organizational service orientation is really crucial for customers' quality perception and business performance. According to some authors organizational service orientation plays a crucial role in the success of enterprises (Homburg et al., 2002; Walker, 2007). Service orientation is positively related to the main service delivery characteristics, and business performance as well. Empirical investigations show the important influence service orientation has on such variables as: service quality image, organizational commitment, and profitability (ROA) in the banking sector (Lytle & Timmerman, 2006). Service orientation is also related to business performance characteristics, such as re-patronage intention and positive word-of-mouth, with the mediating role of staff satisfaction, service value, and customer, whose relationship was demonstrated in the medical service industry (Yoon et al., 2007).

It is conceivable that the issues contained in this concept have a substantial impact on organization–customer interactions, as well as the nature and quality of service delivery (Yoon et al., 2007). Organizational service orientation was also identified as a "common denominator" of educational service attributes that are responsible for clients' satisfaction (Walker, 2007). Nevertheless, in telecommunication call centres the organizational service orientation was identified as a factor that had no influence on service quality, whereas other service climate elements had a significant influence (Little & Dean, 2006). On the other hand, service climate, which is a very similar concept, has been identified as negatively related to the owners' service values (the degree to which owners valued innovativeness, attentiveness, outcome-orientation, aggressiveness, support, and decisiveness) in the small business environment (Andrews & Rogelberg, 2001).

According to Gonzalez & Garazo (2006) the organizational service orientation has a positive influence on employees' satisfaction and organizational citizenship behaviour. Organizational citizenship behaviour was defined by researchers as three main variables: (1) whether employees act as representatives of the firm to outsiders, (2) contact-staff participation consists in providing information about customer needs and suggesting improvements in service delivery process, and (3) following company regulations in such a conscientious manner that they are adapted to the individual customer's needs (Gonzalez & Garazo, 2006). In brief, organizational citizenship behaviour means "go the extra mile" for customers. These are very important elements of excellent service delivery.

The organizational service orientation as a diagnostic approach might also be used in the public services environment. Akesson et al. (2008) proposed areas of service orientation in public e-government services, and their theoretical analyses show that this concept provides a useful contribution to these particular services as well.

5. Similarity to other concepts

According to Vargo & Lusch (2004, p. 11) interactivity, integration, customization and coproduction are the hallmarks of a service-centred approach. Along with these, focus on the customer and the relationship are also very important. In accordance with this thesis several approaches stresses the importance of the contact staff who directly provide the service, and their cultural context. Some authors mention that generally there are two important factors influencing employees' tendencies to provide the quality of service: the first one lies in the organization of a service company, and the second exists in individual personality characteristics (Baydoun et al., 2001; Homburg et al. 2002). The former is described by "macro-organizational approaches", like service climate and service orientation. The latter is the "personality-based approach" and it is focused on the personal skills and other features of staff, who are assessed by psychological tests and similar tools.

Organizational service orientation is often described in the context of service organization climate (Lytle & Timmerman, 2006; Lynn et al., 2000; Lytle et al., 1998). Organizational climate and culture are interconnected. Employees' values and beliefs (part of culture) influence their interpretations of organizational policies, practices, and procedures (climate)

(Schneider, 1996, p. 9). The organizational climate includes employees' perceptions of the policies, practices and procedures that are rewarded, supported, and expected concerning clients (Schneider et al., 2002). The climate of an organization is a summary of employees' impressions about "how we do things around here" or "what we focus on around here" or "what we direct our efforts to around here" (Schneider et al., 2006, p. 117). The climate is the psychological identity of employees in an organization. A climate is researched in the service environment context, and thus it is called service climate (Schneider, 1980; Schneider et al., 2006; Little & Dean, 2006; Walker, 2007).

It can also be observed that the "customer service orientation" concept is in many ways similar to the organizational service orientation, but it is focused on staff behaviours and a more psychological interpretation. Customer service orientation is specified by interpersonal skills, extroversion, and the general disposition of operators having a positive influence on the operators' performance (Alge et al., 2002). It is still perceived as a part of the service climate. Walker (2007) classifies three service climate dimensions as "service orientation"; they are: staff service ethos, staff personal attributes, and staff concern for clients. They were found to be key elements of the organizational service climate. Little & Dean (2006) also classify customer service orientation as a dimension of service climate. They propose four dimensions of service climate, and one of them is customer orientation, which is understood as the degree to which an organization tends to meet customer needs and expectations for service quality.

Baydoun et al. (2001) propose instruments for customer service orientation assessment. These demonstrate the utility of personality variables for predicting service behaviour. Based on this instrument high-quality service providers could be selected. There are more methods for customer service orientation assessment. Martin & Fraser (2002) use the Customer Service Skills Inventory (CSSI) for identification of individuals who are likely to succeed in positions that involve working with the customers or clients of an organization. The CSSI is a short self-report measure of customer service orientation.

The literature also provides the "customer orientation" concept, derived from a relationship marketing approach. The customer orientation concerns service employees who have direct contact with customers. Hennig-Thurau & Thurau (2003) propose customer orientation as a three-dimensional construct: employees' motivation to serve customers, their customer-oriented skills, and self-perceived decision-making authority.

Finally, it is considered that organizational service orientation is part of a wider concept of an organization's overall climate. And it is necessary to admit that the organizational service orientation construct is not clearly defined (Lytle et al., 1998). But it seems to be very important from the point of view of a service firms' development: it mostly concerns an internal organizational system which is created by managers, and it provides a relatively precisely defined field for organizational changes and improvements which aim at service excellence; it might also be useful in organizing monitoring purposes, and in benchmarking others as well.

6. How can organizational service orientation be measured?

Organizational service orientation, as with other corporate culture related concepts, has been measured in many ways and in many service industries last two decades. Researchers have developed several diagnostic tools to identify the state of service orientation. For example, Andaleeb et al., 2007) used the specific survey tool to approximate doctors' service orientation, but in this case the concept of service orientation was understood as a set of doctors' behaviours towards their patients. The established construct of service orientation was more similar to the customer service orientation mentioned above.

A very useful tool for organizational service orientation measurement was proposed by Lytle et al. (1998), and it was named "Serv*Or". Serv*Or consists of 35 question items with Likert's scale. The questionnaire items describe four attributes of organizational service orientation, and these attributes altogether comprise 10 organizational service orientation dimensions. These attributes are contained in the four fields of practices briefly discussed above. These attributes (or dimensions) are as followed: servant leadership, service vision (service leadership practices), customer treatment, employee empowerment (service encounter practices), service failure prevention and recovery, service technology, service standards communication (service system practices), service training, and service rewards (human resource management practices) (Lytle et al., 1998).

The proposed diagnostic tool was tested and validated in the American banking sector and retail builders' suppliers. According to the authors, the Serv*Or tool demonstrates a cross-industry universal instrument for assessing service orientation in other firms, not just banks (Lytle et al., 1998). Authors mention that it can be used across different industries and also different work environments for service orientation diagnosis.

At a later time Serv*Or tool was successfully used several times, inter alia, in the hospitality industry (Gonzalez & Garazo, 2006), medical services (Yoon et al., 2007) and the banking sector (Lynn et al., 2000). Some authors also revised and proposed a modification to the Serv*Or scale. Lee et al. (2001) did this in the case of the hotel industry.

7. Research aims and methods

Despite research experience the organizational services orientation concept seems to be an interesting issue for further and deeper investigation. Nevertheless among evidence showing the positive role of organizational service orientation in service enterprises there are also exceptions. Moreover, we do not at this point have wide, cross-sector studies of this concept going through a large number of service sectors. The proposed research question is whether organizational service orientation is really a predicator of key business performance in the wide scope of service organizations operating in different service sectors. Is it still really important for customers' quality perceptions and for business performance? The organizational service orientation construct can be treated as a part of organizational culture, where specific national aspects play a significant role. One of the first studies on corporate culture stresses noticeable differences in organizations' culture among different countries (Hofstede, 1997). So it also seems interesting to investigate the state and role of organizational service orientation in the economic background of a post-communist country in Central Europe.

Furthermore, the next question is whether organization service orientation differs considerably across service sectors? It is obvious that services across different sectors have a different nature; depending on how close service staff is to customers, if the use of standardization is wide or not, what the roles of capital and human power are, etc.

Investigation of organizational service orientation in a cross-industry context will create enhanced analytical opportunities. Therefore, it is worth identifying which aspects of organizational service orientation are different across the different types of service organizations. This can bring conclusions with reference to the role of service orientation dimensions across service sectors - which of them are equal in different service sectors and which are different in a definite way. Suggestions concerning the service orientation dimensions can be drawn up.

This study proposes the investigation of organizational service orientation in Poland, a Central-Eastern European economy. The empirical investigation was conducted in three regions of Poland: Podlasie, Mazowsze, and Warmia & Mazury. Poland joined the EU in 2004; an adjustment programme lasting a few years had been established before that. After 1989, when the communist system collapsed, the Polish economy changed rapidly. The preparation for joining the EU was a strong improvement impulse for the Polish economy. Nowadays, this economy is in an upward phase (A Study..., 2011). According to the Central Statistical Office in Poland, in 2010 GDP growth amounted to 3.8%, and mostly this was an effect of the increase in individual consumption. Poland was the best performing country in the OECD in 2009, with the economy recording economic growth of 1.7%, and one of the few to avoid recession (Restoring Public Finances, 2011). In fact, Poland does not have huge services participation in comparison with other developed countries, but the service industry has great dynamics.

Serv*Or tool (Lytle et al., 1998) was employed, and a few significant service enterprises' business performance characteristics were also gathered. Organizational service orientation is treated as an independent variable, and service outcome as a dependent variables. In studying the problem of differences of organizational service orientation in service sectors the variable "service sector" is considered as the grouping one for ANOVA. In a deeper investigation of differences between sectors the Least Significant Difference method was employed.

A single enterprise was the research unit. The inquiries were addressed to a manager (or owner if he/she performed a managerial role) from an enterprise. Gonzalez & Garazo (2006) also interviewed managers in the organizational service orientation identification process. The research population comprised 230 service enterprises operating in the three regions of Poland mentioned above, and the research units were chosen randomly. Trained field researchers visited managers in enterprises and asked them questions based on the questionnaire. The Serv*Or question battery was translated and modified, and some of the original research question were combined together so that all these could be clearly understood by respondents. At the same time the managerial language and notions typically used in Polish enterprises were employed. After the preoperational free interviews it was decided to employ the scale 1-5, which seemed to be better for respondents than seven gradual. For the service performance variables identification managerial subjective assessment was taken advantage of.

All the main sectors of the service industry were represented in the research sample (according to EU classification 24 sectors were detached specially for this study). Sectors were not represented equally. The largest ones that appeared in the sample were the construction and building renovation sector, and transport services (both 22 entities) and the smallest two: R&D services (only one firm) and mineral resources exploitation services (also one). In the research sample there were mostly small and medium enterprises; those with fewer than 250 employees constituted 91.5% of the sample.

8. The state of service orientation

According to the above discussion, the organizational service orientation construct shows organization capability to provide excellent service to customers. Therefore, the gathered data will shed light on what extend the companies from the sample are able to offer excellent service to their customers. According to the data, the average score of organizational service orientation comes to 3.56 in the 5-point scale. The statistics are shown in Table 1.

	Mean	Variance	SD
Service orientation	**3.56**	**0.510**	**0.714**
Service leadership practices	**4.02**	**0.922**	**0.960**
Service vision	4.12	1.042	1.021
Servant leadership	4.00	1.299	1.140
Service encounter practices	**3.50**	**0.822**	**0.907**
Customer treatment	4.06	0.526	0.725
Employee empowerment	2.99	1.680	1.296
Service systems practices	**3.46**	**0.683**	**0.826**
Service technology	3.61	1.221	1.105
Service failure prevention	3.71	1.063	1.031
Service failure recovery	3.41	1.080	1.039
Service standards communication	3.39	0.927	0.963
Human resource management practices	**3.22**	**1.128**	**1.062**
Service rewards	3.11	1.248	1.117
Service training	3.39	1.538	1.240

Table 1. Serv*Or variables scores

Going deeply into the attributes of the organizational service orientation it is noticeable that "service leadership practices" are the most highly scored field of this concept in researched enterprises. Scores above four points might be judged as very good. This field contains the strong vision of a service and the stressed role of customers among managers and service staff, as well as managers' personal involvement in service providing process. Leadership is considered by many theories as the primary success factor, and as the basis for building an effective business organization. For example the ISO 9001 standard of quality system also sees leadership as the first basic condition for quality improvement (paragraph 5 of ISO 9001:2008, Requirements for Management).

A very interesting situation is recognised in the field "service encounter practice". On one hand, it is one of the most highly scored variables – "customer treatment", and on the other hand, the attribute "employee empowerment" which is the least scored (and with the largest projection of results; see the variation and the standard deviation). It seems that in the researched enterprises there is a strong focus on providing very good treatment to the customers, but this is done according to standards which are strictly defined in advance. Staff have very little space for free and flexible initiatives. It also suggests that managers prefer having wide control of all operations while serving customers. Nevertheless, leaving so little space, they seriously limit the potential for moments of truth creation. According to Carlson (1987), moments of truth are the essence of superior quality experiences in service.

The system is next to leadership as the primary force standing behind the sustainable growth of any organization, especially in the context of quality improvement. The "service system practices" are performed at a moderate level. It is noticeable that failure prevention practices are particularly important. Considering the organizational service orientation fields level "human resource management practices" seems to be the weakness of the researched companies.

The observations which emerged from the study might be partly explained by the managerial attitudes specific for the country where the research was conducted. Managers in Poland, a country with a heritage of central planning, still prefer to focus on the individual. They consider management success as a single person's achievement, rather than a team success. Hence the importance of strong leadership and very little space for staff empowerment.

9. The link between organizational service orientation and business performance

During interviews with managers the basic data concerning companies' business performance were gathered. First of all, the overall service quality level was identified, as well as changes in enterprises' market share, changes in profitability, client satisfaction and client loyalty. The correlation coefficients were counted between all organizational service orientation fields, and also the global score of the construct with all the above-mentioned performance variables. The results of correlations' calculation are presented in Table 2, with the non-significant (p>0.05) relationships removed.

There are significant relationships between most organizational service orientation fields and business performance variables. However, the values of correlation coefficients are not high. The number of identified significant relationships allows the conclusion that there is an influence of organizational service orientation on many business performance variables. The values of correlation coefficients would not be expected to be as high because of the great diversity of services in the research sample. Correlation analysis proves that organizational service orientation is a very important predictor of service quality performance. So, it could be considered that the organizational service orientation construct is a fairly good measure which can assess the ability of a service organization to provide excellent service. Taking into consideration that many service sectors were examined, it also allows the suggestion that Serv*Or could be a universal cross-sector tool.

A fairly strong influence was observed on overall service quality. Human resource management practices especially emerge as the most influential. This variable has the highest coefficient value. It underlines the role of the effective motivation of staff, as well as the importance of knowledge and training. The next performance variable, which in a significant way is affected by organizational service orientation is client loyalty. The correlation coefficients are significant with all fields, the same as in the case of human resource management practices. The importance of organizational service orientation as a loyalty predicator reinforces the role of direct contact service personnel in clients' loyalty building.

		Gamma	p level
Service orientation	& Quality level	0.331	0.000
Service leadership practices		0.216	0.001
Service encounter practices		0.263	0.000
Service systems practices		0.245	0.000
Human resource management practices		0.308	0.000
Service orientation	& Market share	0.169	0.003
Service systems practices		0.142	0.012
Human resource management practices		0.189	0.001
Service orientation	& Profitability	0.220	0.000
Service leadership practices		0.163	0.006
Service systems practices		0.141	0.011
Human resource management practices		0.231	0.000
Service orientation	& Client satisfaction	0.243	0.000
Service leadership practices		0.271	0.000
Service systems practices		0.173	0.002
Human resource management practices		0.203	0.000
Service orientation	& Client loyalty	0.256	0.000
Service leadership practices		0.237	0.000
Service encounter practices		0.151	0.006
Service systems practices		0.178	0.001
Human resource management practices		0.196	0.000

Table 2. Correlations between organizational service orientation and business performance

We should also pay attention to the fact that there is a field (service encounter practices) which is correlated with loyalty and not correlated with client satisfaction. Satisfaction might be the main loyalty predictor but not often (Oliver, 1999). Nevertheless, loyalty seems to be one of the most important business performance components (Reichheld & Teal, 2001) and the role of service orientation in wining loyal clients seems to be very beneficial.

10. The differences between service sectors

There were 25 categories in the variable "service sector", and one of them was "others". It was decided to employ ANOVA to investigate if there are significant differences in organizational service orientation scores between service sectors. Sectors were not represented equally, and some of them included small entities, and therefore the service sectors including fewer than five entities were rejected from the sample. A one-way ANOVA provides results, as shown in Table 3.

Taking into consideration the variances in different service sectors, there is not a significant difference in the global organizational service orientation score. Also, many of attributes do not show significant differences. But some particular organizational service orientation dimensions vary significantly between sectors. The largest diversity is noticed within "service encounter practices". It seems that service encounter is the most important aspect that diversifies service orientation across sectors. It might be expected that the most remote contact with clients does not require superior organizational service orientation, and that closer and direct relationships require a special kind of service encounter practices. This also concerns services provided to business clients, as this particular kind of relationship requires a different approach in direct contacts. Unfortunately, service sectors were separated using as a basis the Statistical Classification of Economic Activities in the European Community (NACE), which does not allow a clear distinction between B2B and B2C services.

The ANOVA analysis encourages deeper and more detailed explorations related to the service orientation differentials across service sectors. For these purposes Fisher's LSD (Least Significant Difference) procedure should be applied. It allows the investigation of the individual differences between particular variables in pairs of sectors. But in this case it is complicated to trace it in detail because of the great number of pairs (there are hundreds of pairs of sectors and service orientation attributes).

With the support of Statistica software the LSD significances were counted for five organizational service orientation dimensions: those for which simple analysis of variance showed significant differences: service vision, customer treatment, employee empowerment, service standards communication, and service training. For each service sector the numbers of significant pairs were summed up together. In the five organizational service orientation dimensions mentioned above the greatest number of significant pairs of service sectors were identified in "telecommunications and postal services" – altogether 29 pairs, the next one was "construction and renovation services" – altogether 26 pairs, next "vehicle services and petrol retailing" – 19 pairs. Counting the pairs for service orientation variables, most of them were found in "customer treatment" – 55, and "service standards communication" – 46. The LSD output was evaluated and interpreted with great care, having all the time in mind the fact that service sectors are not represented equally in the research sample. The numbers of significant pairs are presented in Table 4 (the classification "others" is excluded, as well as the small numerous sectors).

	SS	df	MS	SS	df	MS	F	p
Service orientation	10.393	18	0.577	78.289	198	0.395	1.460	NS
Service leadership practices	15.006	18	0.834	186.379	199	0.937	0.890	NS
Service vision	36.895	18	2.050	186.958	198	0.944	2.171	0.005
Servant leadership	21.518	18	1.195	258.482	195	1.326	0.902	NS
Service encounter practices	26.498	18	1.472	151.823	199	0.763	1.930	0.015
Customer treatment	22.922	18	1.273	90.756	197	0.461	2.764	0.000
Employee empowerment	61.169	18	3.398	295.826	196	1.509	2.252	0.004
Service systems practices	8.731	18	0.485	142.915	199	0.718	0.675	NS
Service technology	12.364	18	0.687	244.901	189	1.296	0.530	NS
Service failure prevention	15.440	18	0.858	206.008	193	1.067	0.804	NS
Service failure recovery	15.883	18	0.882	216.706	196	1.106	0.798	NS
Service standards communication	27.559	18	1.531	171.805	197	0.872	1.756	0.033
Human resource management practices	25.307	18	1.406	215.557	199	1.083	1.298	NS
Service rewards	29.105	18	1.617	230.411	194	1.188	1.361	NS
Service training	55.293	18	3.072	271.697	197	1.379	2.227	0.004
NS – non-significant (p>0.05)								

Table 3. ANOVA analysis results

First of all we should focus our attention on the telecommunications and postal services, which are, in fact, very specific. There are still state monopolies in a few kinds of service, namely letter delivery; and in phone call services there are only a few strong market players, as in most European countries. It is not a mystery that in most European countries these services very often cause customers to complain; we have also observed action taken by the European Commission prepared to exert change in this market. In the researched country, the telecommunications and postal services sector structure remains largely unchanged from previous years, which influences particular practices in treating customers. In this sector there is rather remote contact between service staff and customers, which surely drives specific organizational service orientation in the variable of "customer treatment", and some others.

The second very interesting sector, which appears to be noticeably different from, this is construction and renovation services. This sector provides services with rather low personal contact with customers. Most of the researched firms provide services as subcontractors on huge building sites, having no, or almost no, contact with the investor, even if he/she is an individual. Taking care of customer and service quality in this sector does not rely on personal interaction to a great extent, but instead it lies in solid manual work and technical support. Perfection in this case is not kindness and the customer's understanding, but in fact it is in doing a professional, timely and robust job. Vehicle repairing and the petrol retailing sector is also characterised by the strong role of service equipment and manual cleverness. Analysing the sectors which differ the most, it might be concluded that service features probably affect the state of organizational service orientation very much, and this explains the differentiation which was identified in the research sample.

	Customer treatment	Employee emp.	Standards comm.	Service vision	Service training	Row sums
Telecommunications and postal services	16	3	6	1	3	29
Construction and renovation services	3	3	7	8	5	26
Vehicle services and petrol retailing	3	-	6	6	4	19
Retailing	2	2	4	8	1	17
Hotel and restaurant industry	2	2	5	2	6	17
Finance and insurance services	7	3	6	1	-	17
Agriculture and wood services	1	3	4	2	-	10
Health and social care services	4	6	-	-	-	10
Business services	3	-	1	2	2	8
Wholesale and commission trading	2	-	4	1	-	7
Transport services	3	-	1	2	1	7
Printing services	2	-	-	1	3	6
IT services	1	4	-	1		6
Waste util., energy and water supply	2	-	-	1	2	5
Services connected with fabrics	1	-	-	2	1	4
Education	2	-	2	-	-	4
Culture and sport services	1	-	-	-	-	1
Sum in columns	55	26	46	38	28	

Table 4. Numbers of significant pairs

LSD analysis has brought one more additional advantage. Thanks to the number of pairs, we were able to recognize "service standards communication" as a variable which is differentiated very much across service sectors (46 significant pairs). Different sectors have their own approaches to communicating service standards, as well as the manners by which they define them. For example in construction work we have a completely different approach to standards spreading than in financial services. The former is based on the individual's knowledge and skills, as well as on the technical control, while the latter is based on the kindness of individual staff, and intensively uses IT technology to define and to execute standards.

11. Conclusions and managerial implications

This study strongly suggests that organizational service orientation plays an important role in achieving business performance by service organizations. Its influence on service quality and client loyalty is substantial, and it also leads to better finance results. Considering that the researched concept is a credible predicator of the quality of service output, it should be used more often in service organizations diagnosis. It might serve managers and consultants as an assessment approach for organizational ability to provide excellent service to clients. Moreover, it might provide a great framework for service organization improvement, acting as a guide for setting up improvement programmes.

The organizational service orientation concept is closely related to the corporate culture concept, but at the same time it is focused only on behaviours (practices) within organizations. And the research has pointed out that it is a powerful approach for investigating how a service organization deals with a service encounter. The concept definitely helps more in the field of practical service organization than in corporate culture recognition. It may be, of course, on one hand as an advantage, but on the other as a weakness.

The concepts of organizational culture or climate might discourage managers because there are so many intangible elements that are difficult for direct observation. But using the Serv*Or instrument, as proposed in this study, it might be beneficial that it does not measure values and beliefs, but it is only focused on practices within the organization. Thanks to this, it is more universal, and it has a potential for use in many cultures and a variety of nations.

The state of organizational service orientation in the researched sample of service providers shows that the weakest element is employee empowerment, which surely originates partly from the national inclination to individualism, and partly from the central planning system that existed in the Polish economy in communist times. On the other hand, there is a very good score in leadership. These observations look coherent because expanded individualism usually fosters strong leadership. The high score in customer treatment proves that among the researched companies strong market orientation is adapted effectively. But these companies do not take full advantage of the possibility of moments-of-truth creation. According to Carlzon (1987) and Gronroos (2007) moments of truth, of course we mean moments of magic, are one of the crucial elements in gaining a superior perceptions of service quality.

The study adopts a cross-sector approach, and thus the diversity between service sectors might be investigated. Wide cross-sector studies are rather a seldom practice in the quality management field, and in this study it was a challenging problem. It was not proved that general indicators of organizational service orientation varied between service sectors, which on the other hands supports the assumption of the universality and wide applicability of the researched concept. However, the calculations allow us to point out some organizational service orientation dimensions that differ noticeably across sectors, i.e. service encounter practices and service standards communication.

There are two important factors that in an appreciable way affect organizational service orientation differentiation between service sectors: the first one, let us call it the "structural factor", relies on the fact that a service orientation is affected by the structure of a sector, like the diversity of companies and competition level; the second one comes from the service providing process characteristics, especially the intensiveness of close and direct relationships between service staff and customers.

All studies have some limitations, and this one has as well. The most important limitation of this study may be found in the fact that the respondents in the researched enterprises were only managers. Their points of view might be different from those of all the staff who are employed in the enterprises. The representation of respondents in particular sectors was also too small, and unfortunately not equal.

12. References

A Study of Poland's Economic Performance in The 1st Quarter of 2011 (2011). Ministry of Economy in Polnad, Warsaw, May 2011.

Akesson, M. ; Skalen, P. &Edvardsson, B. (2008). E-government and service orientation: gaps between theory and practice. *International Journal of Public Sector Management*, Vol.21, No.1, pp. 74-92, ISSN 0951-3558

Alge, B.J. ; Gresham, M.T. ; Heneman, R.L. ; Fox, J. & McMasters, R. (2002). Measuring customer service orientation using a measure of interpersonal skills: a preliminary test in a public service organization. *Journal of Business and Psychology*, Vol.16, No.3, pp. 467-476, ISSN 0889-3268

Andaleeb, S.S. ; Siddiqui, N. & Khandakar, S. (2007). Doctors' service orientation in public, private, and foreign hospitals. *International Journal of Health Care Quality Assurance*, Vol.20, No.3, pp. 253-263, ISSN 0952-6862

Andrews, T.L. & Rogelberg, S.G. (2001). A new look at service climate: its relationship with owner service values in small businesses. *Journal of Business and Psychology*, Vol.16, No.1, pp. 119-131, ISSN 0889-3268

Baydoun, R.; Rose D. & Emperado, T. (2001). Measuring customer service orientation: an examination of the validity of the customer service profile. *Journal of Business and Psychology*, Vol.15, No.4, pp. 605-602, ISSN 0889-3268

Beitelspacher, L.S. ; Richey, R.G. & Reynolds, K.E. (2011) Exploring a new perspective on service efficiency: service culture in retail organizations. *Journal of Services Marketing*, Vol.25, No.3, pp. 215–228, ISSN 0887-6045

Carlzon, J. (1987). *Moments of Truth*, Ballinger Pub., ISBN 0-06-091580-3, Cambridge, USA

Deming, E. (1994). *The New Economincs*, The MIT Press, ISBN 0-262-54116-5, Cambridge, USA

Furrer, O.; Liu, B. & Sudharshan, D. (2000). The relationships between culture and service quality perceptions: basis for cross-cultural market segmentation and resource allocation. *Journal of Service Research*, Vol.2, No.4, pp. 355-71, ISSN 1094-6705

Gebauer, H.; Edvardsson, B. & Bjurko, M. (2010). The impact of service orientation in corporate culture on business performance in manufacturing companies. *Journal of Service Management*, Vol.21, No.2, pp. 237-259, ISSN 1757-5818

Gonzalez, J.V. & Garazo, T.G. (2006). Structural relationships between organizational service orientation, contact employee job satisfaction and citizenship behaviour. *International Journal of Service Industry Management*, Vol.17, No.1, pp. 23-50, ISSN 0956-4233

Gronroos, Ch. (2007). *Service Marketing and Management. Customer Management in Service Competition*, John Wiley & Sons, ISBN 978-0-470-02862-9, Hoboken, USA

Hennig-Thurau, T. & Thurau, C. (2003). Customer Orientation of Service Employees – Toward a Conceptual Framework of a Key Relationship Marketing Construct. *Journal of Relationship Marketing*, Vol.2(1/2), pp. 23-41, ISSN 1533-2667

Heskett, J.L.; Jones, T.O.; Loveman, G.W.; Earl Sasser, Jr.W. & Schlesinger, L.A. (2008). Putting the service-profit chain to work. *Harvard Business Review*, July–August, pp. 118-129, ISSN 0017-8012

Hofstede, G. (1997). *Cultures and Organizations. Software of the Minds,* McGrow-Hill, ISBN 978-0-9742114-7-3, New York, USA

Homburg, C. ; Hoyer, W.D. & Fassnacht, M. (2002). Service Orientation of a Retailer's Business Strategy: Dimensions, Antecedents, and Performance Outcomes. *Journal of Marketing*, Vol.66, pp. 86-101, ISSN 0022-2429

Hsieh, Y.M. & Hsieh, A.T. (2001). Enhancement of service quality with job standardisation. *The Service Industries Journal*, Vol.21, No.3, pp. 147-166, ISSN 0264-2069

Johnston, R., & Clark, G. (2005). *Service operations management,* (2nd Ed.), Prentice-Hall, ISBN 978-0070037939, Harlow, England

Kanousi, A. (2005). An empirical investigation of the role of culture on service recovery expectations. *Managing Service Quality*, Vol.15, No.1, pp. 57-69, ISSN 0960-4529

Kantsperger, R., & Kunz, W.H. (2005). Managing overall service quality in customer care centers. Empirical findings of a multi-perspective approach. *International Journal of Service Industry Management*, Vol.16, No.2, pp. 135-151, ISSN 0956-4233

Lee, Y.K.; Park, D.H. & Yoo, D.K. (2001). The structural relationship between service orientation, mediators and business performance in Korean hotel firms. *Tourism Sciences*, Vol.21, No.1, pp. 49-65, ISSN 1006-575X

Little, M.M. & Dean, A.M. (2006). Links between service climate, employee commitment and employees' service quality capability. *Managing Service Quality*, Vol.16, No.5, pp. 460-476, ISSN 0960-4529

Lynn, M.L. ; Lytle, R.S. & Bobek, S. (2000). Service orientation in transitional markets: does it matter? *European Journal of Marketing*, Vol.34, No.3/4, pp. 279-298, ISSN 0309-0566

Lytle, R.S. & Timmerman, J.E. (2006). Service orientation and performance: an organizational perspective. *Journal of Services Marketing*, Vol.20/2, pp. 136–147, ISSN 0887-6045

Lytle, R.S.; Hom, P.W. & Mokwa, M.P. (1998). Serv*Or: A Managerial Measure of Organizational Service-Orientation. *Journal of Retailing*, Vol.74, Issue4, pp. 455-489. ISSN 0022-4359

Martin, L.A. & Fraser, S.L. (2002). Customer service orientation in managerial and non-managerial employees: an exploratory study. *Journal of Business and Psychology*, Vol.16, No.3, pp. 477-484, ISSN 0889-3268

Mukherjee, A. & Malhotra, N. (2006). Does role clarity explain employee-perceived service quality? A study of antecedents and consequences in call centres. *International Journal of Service Industry Management*, Vol.17, No.5, pp. 444-473, ISSN 0956-4233

Oliver, R.L. (1999). Whence Consumer Loyalty? *Journal of Marketing*, Vol.63, Issue 4 (Special Issue), pp. 33-44, ISSN 0022-2429

Parasuraman, A.; Zeithaml, V.A. & Berry, L.L. (1988). SERVQUAL: A multiple-item scale for measuring consumer perceptions of service quality. *Journal of Retailing*, Vol.64, Issue1, pp. 12-40, ISSN 0022-4359

Peters, T. & Waterman, R.H. (1982). *In Search of Excellence. Lessons from America's Best-Run Companies*, A Warner Communications Company, ISBN 0-446-38281-7, New York, USA

Reichheld, F.F. & Teal, T. (2001). *Loyalty Effect: The Hidden Force Behind Growth, Profits, and Lasting Value*, Harvard Business School Press Books, ISBN 978-1578516872, Boston, USA

Restoring Public Finances (2011). *Special Issue of the OECD Journal on Budgeting*, Vol.2011/2, pp. 1-216, ISSN 1608-7143

Schneider, B. (1980). The Service Organization: Climate Is Crucial. *Organizational Dynamics*, Vol.9, Issue2, pp. 52-65, ISSN 0090-2616

Schneider, B. (1988). Notes on climate and culture, In: Lovelock, C., (Ed.), *Managing Services*, Prentice-Hall, ISBN 0-13-544701-1, Englewood Cliffs, USA

Schneider, B.; Brief, A.P. & Guzzo, R.A. (1996). Creating a Climate and Culture for Sustainable Organizational Change. *Organizational Dynamics*, Vol.24, Issue4, pp. 7-19, ISSN 0090-2616

Schneider, B.; Salvaggio, A. & Subirats, M. (2002). Climate strength: a new direction for climate research. *Journal of Applied Psychology*, Vol.87, pp. 220-229, ISSN 0021-9010

Schneider, B.; Macey, W.H. & Young, S.A. (2006). The Climate for Service: A Review of the Construct with Implications for Achieving CLV Goals. *Journal of Relationship Marketing*, Vol.5, Issue2/3, pp. 111-132, ISSN 1533-2667

Schneider, B.; White, S.S. & Paul, M.C. (1998). Linking service climate and customer perceptions of service quality: Test of a casual model. *Journal of Applied Psychology*, Vol. 83, Issue2, pp. 150-163, ISSN 0021-9010

Specht, N.; Fichtel, S., & Meyer, A. (2007). Perception and attribution of employees' effort and abilities. The impact on customer encounter satisfaction. *International Journal of Service Industry Management*, Vol.18, No.5, pp. 534-554, ISSN 0956-4233

Swanson, S.R. & Kelley, S.W. (2001). Service recovery attributions and word-of-mouth intentions. *European Journal of Marketing*, Vol.35, No.1/2, pp. 194-211, ISSN 0309-0566

Ueno, A. (2008). Is empowerment really a contributory factor to service quality? *The Service Industries Journal*, Vol.28, No.9, pp. 1321–1335, ISSN 0264-2069

Urban, W. (2011). Perceived quality versus quality of processes; a meta concept of service quality measurement. *The Service Industries Journal*, DOI:10.1080/02642069.2011.614337, ISSN 0264-2069

Vargo, S.L. & Lusch, R.F. (2004). Evolving to a New Dominant Logic for Marketing. *Journal of Marketing*, Vol.68 (January), pp. 1-17, ISSN 0022-2429

Walker, J. (2007). Service climate in New Zealand English language centres. *Journal of Educational Administration*, Vol.45, No.3, pp. 315-337, ISSN 0957-8234

Winsted, K.F. (1997). The service experience in two cultures: a behavioral perspective. *Journal of Retailing*, Vol.73, No.3, pp. 337-60, ISSN 0022-4359

Yoon, S.J.; Choi, D.C. & Park, J.W. (2007). Service Orientation: Its Impact on Business Performance in the Medical Service Industry. *The Service Industries Journal*, Vol.27, No.4, pp. 371–388, ISSN 0264-2069

8

Chinese Tourists' Satisfaction with International Shopping Centers: A Case Study of the Taipei 101 Building Shopping Mall

Shu-Mei Wang

Department of Tourism, Shih Hsin University
Taipei, Taiwan

1. Introduction

Shopping is one of the main activities undertaken by tourists (Kent et al., 1983; Choi et al., 1999; Reisinger & Turner, 2002; Snepenger et al., 2003; Kent & Yuksel, 2007). For some tourists, shopping may be the single most important purpose of tourism (Cohen, 1995; Reisinger & Waryzack, 1996; Huang & Hsu, 2005), or be viewed as an indispensable part of being a tourist (Heung & Qu, 1998; Yuksel, 2004). Creating a comfortable, attractive shopping environment is thus a key aspect of tourist industry development (Block et al., 1994; Jones, 1999; Lin, 2004; Yuksel, 2004). Not only can the existence of a first-class shopping environment enhance tourists' overall satisfaction with the tourist destination as a whole (Heung & Cheng, 2000; Yuksel, 2007), it can also encourage tourists to spend more money (Jones, 1999), thereby providing increased economic benefits for the tourist destination (Jansen-Verbeke, 1991; Di Matteo & Di Matteo, 1996; Timothy & Butler, 1995).

According to statistics published by the National Tourism Administration of the People's Republic of China (CNTA), in 2010 there were a total of 57.39 million tourist departures from China, and these Chinese tourists spent a record US$48 billion while traveling overseas. Expenditure on shopping as a percentage of overall tourist expenditure was highest among Chinese tourists traveling to Hong Kong, at 76% of total tourist expenditure (CTA, 2011). Taiwan did not relax the restrictions on travel to Taiwan by Chinese tourists until July 2008, but in the three years that have elapsed since then China has replaced Japan as Taiwan's main source of overseas tourist arrivals, and now accounts for around 50% of all tourists traveling to Taiwan (Tourism Bureau, Taiwan, 2011). Statistics compiled by Taiwan's Tourism Bureau indicate that shopping is the single largest expenditure item for Chinese tourists visiting Taiwan, and that expenditure on shopping by Chinese tourists is higher than average expenditure on shopping by tourists of all nationalities. Given that, as noted above, shopping plays a very important part in tourism, there is a clear need for Taiwan to develop a more in-depth understanding of Chinese tourist shopping satisfaction.

The Taipei 101 Building Shopping Mall is located in the Taipei 101 Building, the second highest building in the world, and one of Taipei's best-known landmarks. According to the results of a survey conducted by Taiwan's National Tourism Administration, the Taipei 101

Building Shopping Mall is Chinese tourists' favorite shopping location within Taiwan. As it is only relatively recently that Taiwan relaxed the restrictions on travel to Taiwan by Chinese tourists, academic research in this area is still largely focused on overall tourism service quality; little effort has been made to explore the shopping service quality aspect of Chinese tourism in Taiwan. Internationally, a large number of studies have been undertaken examining shopping by Chinese tourists, but most of these studies – such as those by Cai et al. (2001), Jang et al. (2003), and Becken (2003) – have focused on the overall economic benefits of shopping by Chinese tourists, or on their consumption behavior and consumption models; few studies have been undertaken of Chinese tourists' shopping satisfaction. Those few studies that have addressed the question of shopping satisfaction – including the studies by Choi et al. (1999), Heung & Cheng (2000), Liu et al. (2008), Tasci & Denizci (2010), and Lee et al. (2011) – have confined the scope of their case studies to Hong Kong or South Korea, and have limited their evaluation to the shopping environment as a whole, with no attempt to appraise shopping satisfaction with respect to individual (or representative) shopping centers. The aim of the present study is to help fill this gap in the empirical literature. It is anticipated that the study will help governments and business enterprises that are hoping to benefit from the consumption power of Chinese tourists to gain a more in-depth understanding of the constitute elements of Chinese tourist shopping satisfaction, thereby helping them to improve service quality in shopping environments catering to Chinese tourists, and boosting the economic benefits from tourism.

2. Review of the literature

2.1 Service packages and service quality

The concept of the "service package" was first used in Fitzsimmons & Fitzsimmons (1994). It is based on the idea that output in the service sector consists of the "package" of goods and services that a company provides within a given environment. Fitzsimmons & Fitzsimmons (2005) expanded the number of service package attributes from four to five: (1) Supporting facilities: The physical resources that must be put in place before a service can be offered; (2) Facilitating goods: The material purchased or consumed by the buyer, or the items provided by the customer; (3) Information: Operations data or information that is provided by the customer to enable efficient and customized service; (4) Explicit service: The benefits that are readily observable by the senses, and that consist of the essential or intrinsic features of the service; (5) Implicit service: Psychological benefits that the customer may sense only vaguely , or the extrinsic features of the service.

Kellogg & Nie (1995) take the service package matrix proposed by Fitzsimmons & Fitzsimmons as their analytical framework; by entering the characteristics of different service industries into the service package, they seek to identify the service characteristics or key features required by individual service industries, to provide a basis for formulating appropriate strategies to enhance service quality. Kandampully (2000) uses the service package concept as a basis for exploring the factors that influence customer satisfaction in the travel industry. This paper suggests that, in all service industries (including the travel industry), the goods and services provided embody a hybrid mix of tangible and intangible elements; customer satisfaction is determined by the customer's experience of, and interaction with, this hybrid mix. Business managers need to simultaneously manage these "hard" and "soft" factors, ensuring that the right balance is kept between them, in order to be able to enhance service quality and raise the level of customer satisfaction. Bloch et al. (1986), Batra & Ahtola (1990), Eroglu et al. (2005), Michon et al. (2005) and Yuksel (2007) all offer confirmation for the idea that "shopping value" derives from two elements: utilitarian value (referring to the consumer's ability to buy goods they need) and hedonic value (referring to the pleasure that the consumer derives from the shopping process). Both types of value can be obtained whenever a consumer undertakes the shopping process Eroglu et al. (2005). In the case of purchasers who are on holiday, hedonic value becomes more significant than it would otherwise be. Snepenger et al. (2003) suggest that shopping allows tourists to experience local culture, and to fit in with local lifestyles. Hughes (1995) and Cary (2004) believe that the shopping process gives tourists a feeling of excitement or novelty. These research results suggest that shopping centers can enhance the enjoyment tourists obtain from shopping through careful service package design; this view is supported by the research of Kellogg & Nie (1995) and Kandampully (2000), which found that identifying the constituent elements of the service package and the optimal service package element mix could help to enhance the quality of service provided by shopping centers.

Heung & Cheng (2000) and Liu et al. (2008) evaluate the indicators used to appraise the shopping satisfaction of Chinese tourists visiting Hong Kong. The study of the shopping behavior of overseas tourists by Dimanche (2003) uses basically the same service quality indicators as those used in previous studies in this area. Yuksel (2004) finds that, while there may be some variation between overseas tourists and locals (i.e. Turkish citizens) when it comes to constitute elements of shopping service quality, the same appraisal indicators can be used for both groups. The present study therefore uses a review of the literature on shopping service quality to identify suitable service quality appraisal indicators. A retrospective examination of past studies of shopping environment service quality shows that the appraisal indicators used all fall broadly within the five service package attributes identified by Fitzsimmons & Fitzsimmons: (1) Supporting facilities: This covers "hardware" infrastructure and facility indicators such as "transportation to the mall," or "cleanliness of shops" (Heung & Cheng, 2000; Yuksel, 2004; Josiam et al., 2005; Liu et al., 2008; Keng et al., 2007); (2) Facilitating goods: This aspect covers the quality of the goods being sold, including the range of products and brands available, the frequency with which new models are introduced, price, reliability, etc. (Heung & Cheng, 2000; Yuksel, 2004; Josiam et al., 2005; Keng et al., 2007; Liu et al., 2008); (3) Information: This includes the language and communication skills of the sales personnel, the presentation of information relating to the shopping facility, and visual merchandising (Heung & Cheng, 2000; Yuksel, 2004; Josiam et al., 2005; Liu et al., 2008); (4) Explicit service: This covers the attitude and professionalism of the sales personnel (Milliman, 1986; Heung & Cheng, 2000; Yuksel, 2004; Josiam et al., 2005; Keng et al., 2007; Liu et al., 2008), waiting time (Hui et al., 1997; Heung & Cheng, 2000), payment methods (Heung & Cheng, 2000), etc.; (5) Implicit service: This aspect includes shopping facility décor, music, color schemes, and scent (Milliman, 1982; Bellizzi et al., 1983; Yalch & Spangenberg, 1990; Bellizzi & Hite, 1992), the aesthetic appeal or special features of the building in which the shopping facility is located (Kotler, 1974; Jansen-Verbeke, 1991; Bitner, 1992; Lin, 2004; Keng et al., 2007), and the sense of excitement, discovery or cultural enlightenment experienced during the travel or shopping process (Hughes, 1995; Snepenger et al., 2003; Cary, 2004).

2.2 Service quality, customer satisfaction and customer loyalty

Most of the studies in the literature have confirmed that service quality constitutes a leading indicator for customer satisfaction (Cronin & Taylor, 1992; Ekinci, 2003; González et al., 2007), and that there is a strong positive correlation between service quality and satisfaction (Taylor & Baker, 1994; Olorunniwo et al., 2006; Hu et al., 2009). Shopping-related research has also shown a significant positive correlation between shopping center service quality and customers' shopping behavior (Kotler, 1974; Babin et al., 2004), and has demonstrated that a high-quality service package can increase the time that customers spend in the shopping center (Spangenberg et al., 1996), as well as boosting the amount of money they spend and the number of items they purchase (Crowley, 1993; Spangenberg et al., 1996), and enhancing customer satisfaction (Mattila & Wirtz, 2001).

A study by Olsen & Johnson (2003) found that product quality, price and service flow combine to affect customer satisfaction. Olsen & Johnson suggest that customer satisfaction derives from the overall transaction experience; their findings thus appear to be in conformity with the service package concept. A business enterprise needs to identify the aspects of the transaction experience that affect customers if they are to be able to make improvements and enhance customer satisfaction, thereby building customer loyalty. In the present study, it is assumed that there is a direct relationship between the constitute elements of service quality (products, price, and service) and satisfaction. Unlike past studies in this field, the present study does not use overall service quality as an intermediary between the constituent elements of service quality and satisfaction; in this respect, the present study is similar to Otto & Ritchie (1996), which treats service quality and satisfaction as being synonymous.

The ultimate goal of improving shopping service quality is to use the provision of a first-class service experience to enhance customer satisfaction, thereby leading the customer to spend more money, and possibly even establishing customer loyalty. Attitudinally speaking, customer loyalty may take the form of brand preference or expression of willingness to purchase (Lee et al., 2001). In behavioral terms, it is expressed by repeat purchasing, and through the building of word-of-mouth brand reputation. Oliver (1999) suggests that satisfaction constitutes an important step in the process of forging loyalty, but that loyalty is not an inevitable result of satisfaction. Berné & Yague (2001) found that those customers who had the highest level of satisfaction did not necessarily wish to engage in re-purchasing. In their study of the level of satisfaction of Chinese tourists visiting Korea et al. (2011) found that Chinese tourists' satisfaction with travel to Korea did not lead to increased loyalty towards travel to Korea. However, Huang & Hsu (2005) came to a different conclusion. They suggested that the fact that Chinese tourists from both Beijing and Guangzhou generally viewed Hong Kong as a "shopper's paradise," and that some survey respondents made regular trips to Hong Kong for shopping, indicated that Chinese tourists did feel loyalty towards travel to and/or shopping in Hong Kong. Overall, it is clear that more research is needed to explore the relationship between satisfaction and loyalty in tourism and shopping.

3. Methodology

3.1 Research model

The purpose of the research model used in the present study is to verify the existence of a causal relationship between the service package offered by the Taipei 101 Building

Shopping Mall and the level of satisfaction and loyalty of Chinese tourists visiting the Mall. The study takes as it foundations the service package theory of Fitzsimmons & Fitzsimmons (2004), while following Otto & Ritchie (1996) and Olsen & Johnson (2003) in viewing service quality as being synonymous with satisfaction. The following hypotheses are made: The five attributes of the service package each affect customer satisfaction (H1 – H5), and customer satisfaction affects customer loyalty (H6) (Fig.1). The five hypotheses established for the individual service package attributes are as follows:

- H1: Supporting facilities have a positive impact on customer satisfaction
- H2: Facilitating goods have a positive impact on customer satisfaction
- H3: Information has a positive impact on customer satisfaction
- H4: Explicit service has a positive impact on customer satisfaction
- H5: Implicit service has a positive impact on customer satisfaction
- H6: Customer satisfaction has a positive impact on customer loyalty

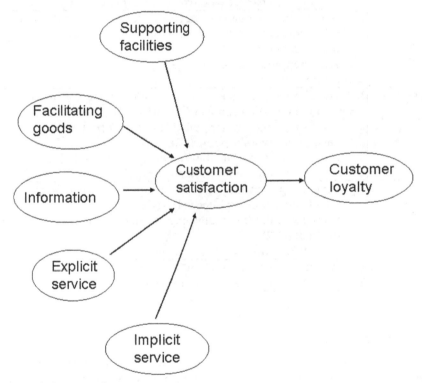

Fig. 1. Research framework

3.2 Questionnaire design

The questionnaire used in the present study was formulated based on the results obtained in the review of the literature. The 20 variables covered by the questionnaire were as shown in Table 1 below.

Variable	Items
Supporting facilities (SF)	1. Convenience of transportation access to the shopping center (SF1) 2. Provision of adequate restrooms, elevators, left luggage lockers, etc. (SF2) 3. Cleanliness of shopping center environment (SF3)
Facilitating goods (FG)	1. Goods sold in the shopping center are reasonably priced (FG1) 2. The shopping center offers a wide range of fashionable products (FG2) 3. The products sold in the shopping center are of reliable quality, and there is no danger of being sold fake/pirated goods (FG3)
Information (IF)	1. Information about the services available on each floor of the shopping center can be obtained easily (IF1) 2. The goods on sale, and circulation within the shopping center, are clearly labeled (IF2) 3. No language barrier when communicating with salespeople (IF3)
Explicit service (ES)	1. Sales personnel display a respectful, friendly, helpful attitude (ES1) 2. Sales personnel demonstrate a high level of knowledge about the products they are selling (ES2) 3. The shopping center offers a wide range of convenient payment methods (ES3)
Implicit service (IS)	1. The shopping center is worth visiting because it is located in the second highest building in the world (IS1) 2. The shopping center provides an opportunity to experience Taiwanese-style shopping (IS2) 3. Having shopped at the Taipei 101 Building Shopping Mall is something you can go home and brag about having done (IS3)
Customer satisfaction (CS)	1. I ended up spending more in the Taipei 101 Building Shopping Mall than I had originally intended to (CS1) 2. Even if you don't buy anything, it's interesting just wandering round the Mall (CS2) 3. Overall satisfaction (CS3)
Customer loyalty (CL)	1. Will come back again (CL1) 2. Would recommend the Mall to others (CL2)

Table 1. Explanation of Variables

4. Survey and analysis

In the present study, a sample survey was administered to Chinese tourists visiting Taiwan; the convenience sampling method was used. Over the period from July 20 to August 5, 2011, pretesting was carried out, using a total of 80 questionnaires, so that any problems could be ironed out. The survey proper was implemented over the period from August 10 to September 15, 2011; the survey was carried out within the Taipei 101 Building Shopping Mall, targeting Chinese tourists aged 18 or over. In all, 350 questionnaires were distributed; 314 valid, completed questionnaires were returned. Analysis of the questionnaire survey results was performed using LISERL 8.0 and SPSS 14.

4.1 Respondents' basic data

The basic data for the respondents were as shown in Table 2 below. The sex ratio was relatively evenly balanced; the vast majority of the respondents were young or middle-aged. The most commonly given occupation was "businessperson," followed by "salaried employee." Respondents with average monthly income in the range of 3,000 – 8,000 Yuan accounted for 54.4% of all respondents.

Characteristics		Frequency	%
Sex	Male	167	53.2
	Female	147	46.8
Age	18 - 24	34	10.8
	25 - 34	62	19.7
	35 - 44	95	30.3
	45 - 54	61	19.4
	55 - 64	33	10.5
	65 or over	29	9.2
Occupation	Public servant	39	12.4
	Salaried employee	100	31.8
	Businessperson	125	39.8
	Farmer	20	6.4
	Other	30	9.6
Monthly Income (CNY)	Under 3,000 Yuan	34	10.9
	3,000 – 5,000 Yuan	72	22.9
	5,000 – 8,000 Yuan	99	31.5
	8,000 – 10,000 Yuan	48	15.3
	Over 10,000 Yuan	61	19.4

Table 2. Respondent Characteristics

4.2 Goodness-of-fit measurement

The next step undertaken after obtaining the questionnaire survey data was to test the reliability of the data. According to Nunnally (1978), Cronbach's Alpha should be greater than 0.7. In the present study, the composite reliability values obtained for each service package attribute were as follows: SF = 0.862; FG = 0.823; IS = 0.923; IF = 0.873; CS = 0.752; CL = 0.743. The composite reliability values for six of the seven items thus conformed to Nunnally's criteria for reliability; while the composite reliability value for ES (0.685) was under the 0.7 threshold, it was very close to the threshold, so the overall reliability of the data was felt to be acceptable.

4.3 Relationship results

In this section, we begin by evaluating the Goodness-of-Fit Index (GFI) for the model as a whole, in order to confirm that the model established for the present study has good explanatory power with respect to the data collected. Comparison with the appraisal criteria and recommended values proposed in other studies suggests that the model adopted in the present study does in fact have good explanatory power. Joreskog & Sorbom (1989) suggest

Construct	Measure	Standardized Loading	t-value	Measurement Error	Composite Reliability	Average Variance Extracted
SF	SF1	0.27	15.12	0.05	0.862	0.693
	SF2	0.53	17.91	0.11		
	SF3	0.57	17.63	0.14		
FG	FG1	0.59	16.29	0.19	0.823	0.618
	FG2	0.41	16.43	0.07		
	FG3	0.50	14.84	0.19		
IF	IF1	0.56	16.32	0.18	0.873	0.710
	IF2	0.33	14.49	0.09		
	IF3	0.66	20.20	0.08		
ES	ES1	0.50	11.93	0.33	0.685	0.420
	ES2	0.57	12.08	0.42		
	ES3	0.51	11.25	0.40		
IS	IS1	0.70	24.67	0.01	0.923	0.801
	IS2	0.58	24.60	0.01		
	IS3	0.71	14.88	0.31		
CS	CS1	0.56	10.96	0.60	0.752	0.503
	CS2	0.57	14.65	0.28		
	CS3	0.61	18.72	0.12		
CL	CL1	0.44	12.03	0.29	0.743	0.596
	CL2	0.58	18.78	0.07		

Table 3. Measurement Properties of Variables

that the GFI and the Adjusted Goodness-of-Fit Index (AGFI) should fall within the range of 0.8 – 0.89. Mulaik, James, Altine, Bennett, Lind & Stilwell (1989) propose that a satisfactory model should have a Parsimony Goodness-of-Fit Index (PGFI) value of at least 0.5. For the model adopted in the present study, GFI = 0.88, AGFI = 0.84, and PGFI = 0.65; all of these values fall within an acceptable range. Hair, Tatham, Anderson & Black (1998) hold that the Normed Fit Index (NFI) should be greater than 0.9, while Bentler & Bonnett (1980) suggest that a Non-Normed Fit Index (NNFI) value of 0.9 indicates very high goodness of fit; the Comparative Fit Index (CFI) should ideally be greater than 0.9 (Bentler, 1990). The values obtained for the model used in the present study are: CFI = 0.95; NFI = 0.92; NNFI = 0.93; all of these values indicate high goodness of fit. Byrne (1989) suggests that the Root Mean Square Residual (RMR) value should be less than 0.05, while Hu & Bentler (1999) hold that the Standardized RMR (SRMR) should be lower than 0.08. The RMR and SRMR values obtained for the present study are both far below the respective levels noted above. It can thus be seen that, for the model used in the present study, all of the goodness-of-fit indicators fall within the acceptable range, indicating a high level of goodness of fit.

The structural model estimation results are shown in Table 4 and Figure 2 below. The test results obtained for each of the study's hypotheses were as follows:

1. H1 is not supported; supporting facilities do not have a significant impact on shopping center customers' level of satisfaction
2. H2 is supported; facilitating goods have a direct, positive impact on shopping center customers' level of satisfaction (with a value of 0.31, the path coefficient is significant)

3. H3 is not supported; information does not have a significant impact on shopping center customers' level of satisfaction

4. H4 is not supported; explicit service does not have a significant impact on shopping center customers' level of satisfaction

5. H5 is supported; explicit service has a direct, positive impact on shopping center customers' level of satisfaction (with a value of 0.10, the path coefficient value is significant)

6. H6 is supported; customer satisfaction has a direct, positive impact on customer loyalty (with a value of 0.88, the path coefficient value is significant)

Research hypothesis	Path-coefficients	t-Value	Results of analysis
H1: Supporting facilities➔ Custom satisfaction	0.07	0.41	Not supported
H2: Facilitating goods ➔ Custom satisfaction	0.31	1.9	Supported
H3: Information ➔ Custom satisfaction	0.08	0.79	Not supported
H4: Explicit service ➔ Custom satisfaction	0.32	1.47	Not supported
H5: Implicit service ➔ Custom satisfaction	0.1	1.88	supported
H6: Custom satisfaction➔ Custom loyalty	0.88	19.59	Supported

**t >1.96, *t> 1.64

Table 4. Research Hypothesis Results

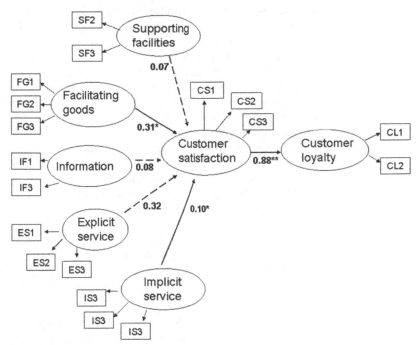

Fig. 2. Structural equation model test results

5. Discussion of the results

The empirical results obtained in the present study show that the level of satisfaction among Chinese tourists with respect to the Taipei 101 Building Shopping Mall is relatively high. The main sources of this satisfaction are the "facilitating goods" and "implicit service" attributes of the service package. Facilitating goods fall under the category of utilitarian value, while implicit service falls under the category of hedonic value. However, in the case of the Taipei 101 Building Shopping Mall, the importance of facilitating goods is greater than that of implicit service. This suggests that the value that Chinese tourists derive from shopping while engaged in tourism is mainly utilitarian value, a finding that goes against the view expressed in most studies in the literature that the pleasure which tourists gain from shopping while engaged in tourism derives mainly from hedonic value. It would thus appear that the constituent elements of the satisfaction that Chinese tourists derive from shopping are different from those that apply in the case of tourists of other nationalities.

Facilitating goods are the aspect of service quality with which Chinese tourists are most concerned, and also their main source of satisfaction. This probably relates to the disparities between the consumption environment in China and that existing in other parts of the world. High import duty makes imported goods significantly more expensive in China than they are in other countries; Chinese tourists therefore like to take advantage of the opportunity to purchase goods at lower prices when they are traveling overseas, and Chinese citizens who do not have the chance to travel overseas will ask friends or relatives traveling abroad to shop on their behalf. This situation is one of the key factors behind the large sums of money that Chinese tourists spend while on holiday. With regard to the three items about which Chinese tourists were asked under the facilitating goods category – "goods sold in the shopping center are reasonably priced," "the shopping center offers a wide range of fashionable products," and "the products sold in the shopping center are of reliable quality, and there is no danger of being sold fake/pirated goods," "goods sold in the shopping center are reasonably priced" had the highest loading value, which is in conformity with the analysis presented above. "The shopping center offers a wide range of fashionable products" had the lowest loading value, possibly because it is only relatively recently that Taiwan has begun to allow Chinese tourists to visit the country, and a high percentage of the Chinese tourists visiting Taiwan will previously have been to "shoppers' paradises" such as Paris and Hong Kong; the range of products and brands on sale at the Taipei 101 Building Shopping Mall may seem relatively unimpressive by comparison, but the prices are still low enough (compared with those charged in China) to ensure that Chinese tourists feel satisfaction after shopping at the Taipei 101 Building Shopping Mall. A similar situation can be seen in Chinese tourist shopping activity in Hong Kong and in the U.S. (Liu, Choi & Lee, 2008; Xu & McGehee, 2011). "The products sold in the shopping center are of reliable quality, and there is no danger of being sold fake/pirated goods" had a somewhat higher loading value, which probably reflects the inadequate enforcement of intellectual property rights protection in China, as a result of which the possibility of unwittingly buying fake or pirated goods in China is quite high, a risk that does not exist when shipping at the Taipei 101 Building Shopping Mall.

Of the three items included under the implicit service attribute, the relatively high loading values of "the shopping center is worth visiting because it is located in the second highest building in the world" and "the shopping center provides an opportunity to experience

Taiwanese-style shopping" are in conformity with the frequent reports in the literature that the opportunity to explore new things and interact with local people is an important aspect of the shopping experience for many tourists. The reaction to "having shopped at the Taipei 101 Building Shopping Mall is something you can go home and brag having done" probably reflects the importance of displaying one's wealth in contemporary Chinese society, an importance seen, for example, in the fact that Swiss watches are sold (using Chinese-language labeling and advertising) on the viewing platform of the Jungfraujoch in Switzerland. Chinese tourists want to do more than just buy the world's most prestigious luxury goods; they want to purchase those goods at famous global locations, in order to maximize the "display of wealth" effect. Given this cultural trait, it is clear that, within the value scheme of Chinese tourists, the Taipei 101 Building Shopping Mall is more than just a world-famous building; it is also an important location for the sale of international luxury brands.

There are two possible reasons why the results obtained for supporting facilities do not indicate any significant impact on customer satisfaction. Firstly, there is the fact that Chinese tourists are currently only allowed to visit Taiwan on package tours. As such, they arrive at the Taipei 101 Building Shopping Mall on tour buses as part of a tour group; they do not need to make their way to the Mall by themselves, and therefore they have no need to concern themselves with the questions of whether transportation access is convenient, whether there are sufficient parking space, or whether enough storage lockers have been provided. Furthermore, package tours usually have a busy itinerary, spending only a limited amount of time at each stop. Chinese tourists on package tours spend an average of only around 2.5 hours at the Taipei 101 Building Shopping Mall, and they spend most of this time shopping or taking souvenir photographs. Provided that there are no major problems, the details of the shopping environment at the Taipei 101 Building Shopping Mall are unlikely to have much impact on Chinese tourists visiting the Mall.

As regards the information attribute of the service package, this attribute does not appear to have much impact on the satisfaction of Chinese tourists visiting the Taipei 101 Building Shopping Mall, possibly because of the lack of a significant language barrier between Chinese and Taiwanese people. However, shopping centers in non-Chinese-language environments might need to take this factor into account. Chinese tourists who have visited the U.S. report that the language barrier spoiled their enjoyment of the shopping experience; Chinese tourists expressed the hope that shopping centers in the U.S. could recruit more Chinese-speaking sales staff and provide more Chinese-language signage and other materials (Xu & McGehee, 2011). Taiwan has an inherent advantage when it comes to the language environment; however, this advantage is gradually being eroded. Shopping centers in Europe, Hong Kong and Japan are all able to recruit sales personnel who speak fluent English, and are starting to provide more in the way of Chinese-language information; by creating a more Chinese-friendly shopping environment, shopping centers in these countries are seeking to develop the significant business opportunities presented by Chinese tourists.

Rather unexpectedly, the explicit service attribute did not appear to have a significant impact on the shopping satisfaction experienced by Chinese tourists at the Taipei 101 Building Shopping Mall. Bearing in mind that the products on sale at the Taipei 101 Building Shopping Mall are mainly international luxury goods, and that luxury goods retailers generally emphasize the creation of a "personal" shopping experience, in which sales assistants give individual explanations of the special features of the products being sold, so as to emphasize the brand value, and given the non-existence of a language barrier, it would seem only natural that tourists shopping at the Taipei 101 Building Shopping Mall are benefiting from explicit service. One possible explanation for the fact that the explicit service attribute did not have a significant impact on customer satisfaction is the short time that Chinese tourists spend at the Taipei 101 Building Shopping Mall; it may be that they simply do not have enough time to enjoy the explicit value provided (in terms of listening to product explanations, trying clothes on, experimenting with different color combinations, etc.). If these aspects of service provision are eliminated, then all that it left is the exchange of physical goods for money. This would explain why the facilitating goods attribute was the main source of satisfaction within the overall service package. Another possible explanation for the fact that explicit service is not a significant contributor to Chinese tourists' shopping satisfaction may be the restrictions that the Chinese government imposes on opportunities for Chinese citizens to travel overseas; those Chinese who are in a position to travel abroad are people towards the top of the social pyramid. People of this sort will be familiar with leading global brands, and will have obtained all the product information they need before leaving China; they will not require explanations from luxury goods boutique sales personnel. Evidence for this is provided by the fact that Chinese tourists can often be seen standing in front of the counters in boutiques in the Taipei 101 Building Shopping Mall holding shopping lists; having clearly already planned out in advance what they are going to buy, they do not require any special assistance from the sales staff.

The shopping satisfaction experienced by Chinese tourists visiting the Taipei 101 Building Shopping Mall has a pronounced, direct impact on customer loyalty. The survey results show that, not only are Chinese tourists very satisfied with their shopping experience in the Taipei 101 Building Shopping Mall, they would also be willing to recommend the Taipei 101 Building Shopping Mall to other people after returning home to China. This is an important finding, because, when they are planning trips aboard, Chinese people are usually strongly influenced by the recommendations they receive from relatives and friends (Beerli & Martin, 2004; Sparks & Pan, 2009); in other words, word-of-mouth marketing is very effective in China. With the first FIT Chinese tourists (i.e. tourists traveling individually rather than as part of a tour group) starting to arrive in Taiwan on June 28, 2011, it can be anticipated that many Chinese tourists who decide to visit Taiwan again will return to the Taipei 101 Building Shopping Mall. With the right government policies in place, the high level of satisfaction that Chinese tourists feel with their shopping experience in the Taipei 101 Building Shopping Mall can be translated into repeat visits and additional purchasing. Given that the number of Chinese tourists that have visited the Taipei 101 Building Shopping Mall already exceeds 2.4 million (according to data compiled by Taiwan's Tourism Bureau in 2011), the potential benefits from the customer loyalty being built up among Chinese tourists towards the Taipei 101 Building Shopping Mall are significant.

6. Conclusions and recommendations

Currently, facilitating goods constitute the main source of shopping satisfaction for Chinese tourists visiting the Taipei 101 Building Shopping Mall. The key factor here is the disparity between the price of luxury goods in Taiwan and China. However, given that China has adopted a policy of trade liberalization, it can be anticipated that import duty in China will gradually be reduced in the future, possibly causing the relative price of imported luxury goods to fall; in this case, the Taipei 101 Building Shopping Mall's price advantage would be eroded. At the same time, with more and more foreign brands opening direct sales outlets in China, the risk of unwittingly buying pirated or fake goods in China is starting to fall. Given these potential changes in the wider business environment, the Taipei 101 Building Shopping Mall may need to think about diversifying the range of products on sale in the Mall, or selling products that are more distinctive to the Mall, so as to be able to maintain its advantage in terms of facilitating goods. This might involve collaborating with leading international brands on the development of limited edition, "Taipei 101" co-branded goods, so as gain maximum benefit from the general feeling among Chinese tourists that the Taipei 101 Building is a major global landmark (as well as being a place to purchase international luxury goods). By leveraging Chinese tourists' desire to show off their wealth, the Taipei 101 Building Shopping Mall should be able to transform itself from a distribution channel into a brand in its own right, leveraging its fame as a landmark to create additional sales revenue.

Following the relaxation in 2003 by the Chinese government of the restrictions on travel to Hong Kong by residents of other parts of China, Hong Kong's tourism and shopping related industries experienced dramatic growth (Tasci & Denizci, 2010). The Chinese government began to permit FIT tourism to Taiwan by individual Chinese travelers starting from June 2011. Although this policy currently applies only to residents of Beijing, Shanghai and Xiamen, it can be anticipated that the scope of implementation will be expanded in the future, creating the possibility that Taiwan could become a major destination for Chinese FIT travel, along the same lines as Hong Kong, and possibly also a major overseas shopping destination. In light of the high level of satisfaction that Chinese tourists have expressed with their shopping experience at the Taipei 101 Building Shopping Mall, and their reported strong interest in coming back again (and in recommending the Mall to others), and given that land and labor costs in Taiwan are significantly lower than in Hong Kong, Taiwan should be able to compete effectively against Hong Kong on price in this respect. The results of a survey of Chinese tourists conducted by Liu, Choi & Lee (2008) showed that the areas where Chinese tourists felt that most improvement was needed in the Hong Kong shopping environment included: more variety of goods and styles; more convenient transportation, better store display design, and the creation of a better in-store atmosphere. The first two of these items fall under the "facilitating goods" attribute, which currently constitutes the core service element for the Taipei 101 Building Shopping Mall; these survey results therefore support the idea that the Taipei 101 Building Shopping Mall has the potential to be highly competitive in the future. The latter two items where Chinese tourists visiting Hong Kong felt improvement was needed fall under the "supporting facilities" and "implicit service" attributes. At present, with most of the Chinese tourists who visit the Taipei 101 Building Shopping Mall doing so as members of tour groups, these attributes may not be particularly important, but once Chinese FIT travel to Taiwan takes off, they can be expected to emerge as important factors influencing the level of satisfaction with shopping services. This is a challenge that the Taipei 101 Building Shopping Mall will be facing in the very near future. The Taipei 101 Building Shopping Mall should try not merely to avoid displaying the deficiencies in service quality that Chinese tourists have reported in Hong Kong, but rather to build competitive advantage in these areas, so as to be able to maintain the loyalty of existing customers and continue to develop new business opportunities.

The research model used in the present study relied heavily on the results of past research in this area as reported in the literature; the study can thus be seen as falling under the category of "verificatory research." However, the service quality attributes that have traditional been used in research on shopping centers do not appear to be fully able to explain the overseas shopping behavior of Chinese tourists, or the component elements of service quality. Comparison with case studies of Hong Kong suggests that there may be significant disparities in shopping service satisfaction preferences between tourists on package tours and FIT tourists. Furthermore, it would appear that, due to the restrictions that the Chinese government imposes on overseas travel, as result of which those Chinese able to travel abroad are generally people with economic and social status that locates them towards the top of the social pyramid, and because of the distinctive social value schemes and consumption habits that have emerged as a result of China's rapid economic development, the research assumptions that need to be made when studying Chinese tourists are different from those which have been applied in past research on tourists from other countries. When considering the huge numbers and immense buying power of Chinese tourists traveling overseas, from the point of view of a researcher studying service quality management, it would seem advisable to adopt an exploratory research approach, with the aim of formulating a "shopping service quality scale" applicable to Chinese tourists that can serve as a management tool to help countries and business enterprises that are seeking to develop the business opportunities presented by Chinese tourists to manage service quality more effectively.

7. References

Babin, J. B., Chebat, J., & Michon, R. (2004). Perceived appropriateness and its effect on quality, affect and behaviour. *Journal of Retailing and Consumer Services*, Vol.11, pp. 287-298, ISSN 0959-0552

Batra, R., & Ahtola, O. (1990). Measuring the hedonic and utilitarian sources of consumer attitudes. *Marketing Letters*, Vol.2, No.2, pp. 159-170, ISSN 0923-0645

Becken, S. (December 2003). Chinese tourism to New Zealand, In: *Landcare Research*, 16.10.2011, Available from <http://www.landcareresearch.co.nz/ research/sustainablesoc/ tourism/ documents /Chinese_tourism.pdf>.

Beerli, A., & Martin, J.D. (2004). Factors influencing destination image. *Annals of Tourism Research*, Vol. 1, No.3, pp. 657-681, ISSN 0160-7383

Bellizzi, A. J., Crowley, E. A., & Hasty, W. R. (1983). The effects of colour in store design. *Journal of Retailing*, Vol.59, No.1, pp. 21-45, ISSN 0022-4359

Bellizzi, J. A., & Hite, R. A. (1992). Environmental colour, consumer feelings and purchase likelihood. *Psychology and Marketing*, Vol. 9, No. 3, pp. 347-363, ISSN 1520-6793

Bentler, P. M., & Bonnett, D.G. (1980). Significant tests and goodness of fit in the analysis of covariance structure. *Psychological Bulletin*, Vol..88, No.30, pp. 588-606, ISSN 0033-2909

Berné, J. M. M., & Yague, M. J. (2001). The effect of variety-seeking on customer retention in services. *Journal of Retailing & Consumer Services*, Vol. 8, pp. 335-345, ISSN 0969-6989.

Bitner, M. (1992). Servicescapes: The impact of physical surroundings on customers and employees. *Journal of Marketing*, Vol. 56, pp. 57-71, ISSN 0022-2429

Bloch, H. P., Sherrell, D., & Ridgway, N. (1986). Consumer search: An extended framework. *Journal of Consumer Research*, Vol.13, pp. 119-126, ISSN 0093-5301

Bloch, H. P., Ridgway, M. N., & Dawson, A. S. (1994). The shopping mall as a consumer habitat. *Journal of Retailing*, Vol.70, No.1, pp. 23-42, ISSN 0022-4359

Byrne, B. M. (1989). *A primer of LISREL: Basic applications and programming for confirmatory factor analytic models*. ISBN 10: 0387969721, New York: Springer-Verlag.

Cai, L.A., Lehto, X.Y., & O'Leary, J. T. (2001). Profiling the U.S.-bound Chinese travelers by purpose of trip. *Journal of Hospitality and Leisure Marketing*, Vol.7, pp. 3-16, ISSN 1050-7051

Cary, S.H.(2004), The tourist moment. *Annals of Tourism Research*, Vol.31, No.1, pp. 61-77, ISSN 0160-7383

China Tourism Academy (2011). *CHINA'S tourism performance review and forecast (2010~2011)*, ISBN 978-7-5032-4106-2, Beijing: Chinese Traveling Publishing house.

Choi, W. M., Chan, A., & Wu, J. (1999). A qualitative & quantitative assessment of Hong Kong's image as a tourist destination. *Tourism Management*, Vol.20, No.3, pp. 361-365, ISSN 0261-5177

Cohen, E. (1995). Touristic Craft Ribbon Development in Thailand. *Tourism Management*, Vol.16, No.3, pp. 225-35, ISSN 0261-5177

Cronin, J. J. & Taylor, S. A. (1992). Measuring service quality: a reexamination and extension. *Journal of Marketing*, Vol.56, No.3, pp. 55-68, ISSN 0022-2429

Crowley, A. E. (1993). The two dimensional impact of color on shopping. *Marketing Letters*, Vol. 4, No.1, pp. 59–69, ISSN 0923-0645

Di Matteo, L., & Di Matteo, R. (1996). An analysis of Canadian cross-border travel. *Annals of Tourism Research*, Vol.23, No.1, pp.103-122, ISSN 0160-7383

Dimanche, F. (2003). The louisiana tax free shopping program for international visitors: a case study. *Journal of Travel Research*, Vol.41, No.2, pp. 311-314, ISSN 0047-2875

Ekinci, Y.(2003) An investigation of the determinants of customer satisfaction. *Tourism Analysis*, Vol.8, No.2, pp.193-196, ISSN 1083-5423

Eroglu, S. A., Machleit, K. & Barr, T. F. (2005). Perceived retail crowding and shopping satisfaction: The role of shopping values. *Journal of Business Research*, Vol.58, No.8, pp. 1146-1153, ISSN 0148-2963

Fitzsimmons, J. A. & Fitzsimmons, M. J.(1994). *Service Management: Operations, Strategy, and Information Technology*(2nd ed.). ISBN 0-07-021760-2 New York: McGraw-Hill.

Fitzsimmons, J. A. & Fitzsimmons, M. J.(2004). *Service Management: Operations, Strategy, and Information Technology*(4th ed.). ISBN : 0-07-282373-9, New York: McGraw-Hill.

González, M. E. A., Comesaña, L. R. & Brea, J. A. F. (2007). Assessing tourist behavioral intentions through perceived service quality and customer satisfaction. *Journal of Business Research*, Vol.60, No.2, pp.153-160, ISSN 0148-2963

Hair, J.F., Tatham, R.L., Anderson, R.E. & Black, W., (1998). *Multivariate Data Analysis* (5th ed.). ISBN 0-13-894858-5, New Jersey: Prentice Hall.

Heung, V. C. S., & Qu, H. (1998). Tourism shopping and its contributions to Hong Kong. *Tourism Management*, Vol. 19, No. 4, pp.383-386, ISSN 0261-5177

Heung, V. C. S. & Cheng, E.(2000). Assessing Tourists' Satisfaction with Shopping in the Hong Kong Special Administrative Region of China. *Journal of Travel Research*, Vol.38, No.4, pp. 396-404, ISSN 0047-2875

Hu, H.H.; Kandampully, J.; Juwaheer, T. D. (2009). Relationships and impacts of service quality, perceived value, customer satisfaction, and image: an empirical study. *The Service Industries Journal*, Vol. 29, No. 2, pp. 111-125, ISSN 0264-2069

Hu, L.T. & Bentler, P. M. (1999). Cutoff criteria for fit indexes in covariance structure analysis: Coventional criteria versus new alternatives, *Structural Equation Modeling*, Vol.6, No.1, pp. 1-55, ISSN 1070-5511

Huang, S. & Hsu, C.H.C. (2005). Mainland Chinese residents' perceptions and motivations of visiting Hong Kong: Evidence from focus group interviews. *Asia Pacific Journal of Tourism Research*, Vol.10, No.2, pp. 191-205, ISSN 1094-1665

Hughes, G.(1995) Authenticity in tourism. *Annals of Tourism Research*, Vol.22, No.4, pp. 781-803, ISSN 0160-7383.

Hui, M. K., Dube, L., & Chebat, J. (1997). The impact of music on consumers' reactions to waiting for services. *Journal of Retailing*, Vol.73, No.1, pp. 87-104, ISSN 0022-4359

Jang, S., Yu, L. & Pearson, T. E. (2003). Chinese travelers to the United States: a comparison of business travel and visiting friends and relatives. *Tourism Geographies*, Vol.5, No.1, pp. 87-108, ISSN 1461-6688

Jansen-Verbeke, M. (1991). Leisure shopping: A magic concept for the tourism industry? *Tourism Management*, Vol.12, No.1, pp. 9-14, ISSN 0261-5177

Jones, M. (1999). Entertaining shopping experiences: An exploratory investigation. *Journal of Retailing and Consumer Services*, Vol.6, No.3, pp. 129-139, ISSN 0959-0552

Joreskog, K. G. & D. Sorbom (1989). *LISREL 7 : A Guide to Program and Application* (2nd ed.). ISBN: 013-177965-6, Chicago : SPSS, Inc.

Josiam, B. M., Kinley, T. R., & Kim, Y. K. (2005). Involvement and the tourist shopper: Using the involvement construct to segment the American tourist shopper at the mall. *Journal of Vacation Marketing*, Vol.11, No.2, pp. 135-154, ISSN 1356-7667

Kandampully, J. (2000). The impact of demand fluctuation on the quality of service: a tourism industry example. *Managing Service Quality*, Vol.10, No.1, pp. 10-18, ISSN 0960-4529

Keng, C. J., Huang, T. L., Zheng, L.J., & Hsu, M. K. (2007). Modeling service encounters and customer experiential value in retailing. *International Journal of Service Industry Management*, Vol.18, No.4, pp. 349-367, ISSN 0956-4233

Kent, W., Schock, P., & Snow, R. (1983). Shopping:Tourism' s Unsung Hero(ine). *Journal of Travel Research*, Vol.21, No.4, pp. 2-4, ISSN 0047-2875

Kellogg, D. L.& Nie, W.,(1995). A framework for strategic service management. *Journal of Operations Management*, Vol.13, No.4, pp. 323-337, ISSN 0272-6963

Kotler, P. (1974). Atmospherics as a marketing tool. *Journal of Retailing*, Vol.49, No.4, pp. 48-61, ISSN 0022-4359

Lee, J., Lee, J., & Feick, L. (2001). The impact of switching costs on the customer satisfaction-loyalty link: mobile phone service in France. *Journal of Services Marketing*, Vol.15, No.1, pp. 35-48, ISSN 0887-6045

Lee, S. Jeon, S. & Kim, D. (2011). The impact of tour quality and tourist satisfaction on tourist loyalty: The case of Chinese tourists in Korea. *Tourism Management*, Vol.32, No.5, pp. 1115-1124, ISSN 0261-5177

Lin, Y. I. (2004). Evaluating a servicescape: The effect of cognition and emotion. *International Journal of Hospitality Management*, Vol.23, No.2, pp.163-178, ISSN 0278-4319

Liu, S. C., Choi, T. M. & Lee, W. T. (2008) Tourists' satisfaction levels & shopping preferences under the solo travel policy in Hong Kong. *Journal of Fashion Marketing and Management*, Vol.12, No.3, pp. 351-364, ISSN 1361-2026

Mattila, S. A., & Wirtz, J. (2001). Congruency of scent and music as a driver of in-store evaluations and behaviour. *Journal of Retailing*, Vol.77, pp. 273-289, ISSN 0022-4359

Michon, R., Chebat, J., & Turley, L. W. (2005). Mall atmospherics: The interaction effects of the mall environment on shopping behaviour. *Journal of Business Research*, Vol.58, No.5, pp.576-583, ISSN 0148-2963

Milliman, R. E. (1982). Using background music to affect the behavior of supermarket shoppers. *Journal of Marketing*, Vol.46, No.3, pp.286-289, ISSN: 0022-2429

Milliman, R. E. (1986). The influence of background music on the behaviour of restaurant patrons. *Journal of Consumer Research*, Vol.13, No.2, pp.86-91, ISSN 0093-5301

Mulaik, S. A., James, L. R., Altine, J. V., Bennett, N., Lind, S., & Stilwell, C. D. (1989). Evaluation of goodness-of-fit indices for structural equation models. *Psychological Bulletin*, Vol.105, No.3, pp. 430-445, ISSN 0033-2909

Nunnally, J. C. (1978). *Psychometric theory* (2nd ed.). ISBN 10: 007047849X. New York: McGraw-Hill.

Oliver, R.L. (1999). Whence Consumer Loyalty? *Journal of Marketing*, Vol.63, pp.33-44, ISSN 0022-2429

Olorunniwo, F., Hsu, M. K. & Udo, G. J. (2006). Service Quality, Customer Satisfaction, and Behavioral Intentions in the Service Factory. *Journal of Services Marketing*, Vol.20, No.1, pp.59-72, ISSN 0887-6045

Olsen, L. L. & Johnson, M. D. (2003). Service equity, satisfaction, and loyalty: from transaction-specific to cumulative evaluation. *Journal of Service Research*, Vol.5, No. 3, pp. 184-195, ISSN 1094-6705

Otto, J. E., Ritchie, J. R. B. (1996). The service experience in tourism. *Tourism Management*, Vol.17, No.3, pp. 165-174, ISSN 0261-5177

Reisinger, Y. & Waryszak, R. (1996). Catering to Japanese tourists: What service do they expect from food and drinking establishments Australia? *Journal of Restaurant and Foodservice Marketing*, Vol.1, No.3/4, pp.53-71, ISSN 1052-214X

Reisinger, Y. & Turner, L. W.(2002). The Determination of Shopping Satisfaction of Japanese Tourists Visiting Hawaii and the Gold Coast Compared. *Journal of Travel Research*, Vol.41, No.2, pp.167-176, ISSN 0047-2875

Snepenger, J. D., Murphy, L., O' Connell, R., & Gregg, E. (2003). Tourists and residents use of a shopping space. *Annals of Tourism Research*, Vol.30, No.3, pp.567-580, ISSN 0160-7383

Spangenberg, E. R., Crowley, A. E., & Henderson, P. W. (1996). Improving the store environment: Do olfactory cues affect evaluations and behaviours. *Journal of Marketing*, Vol.60, pp. 67-89, ISSN 0022-2429

Sparks, B., & Pan, G. W. (2009). Chinese Outbound tourists: understanding their attitudes, constraints and use of information sources. *Tourism Management*, Vol.30, No.4, pp. 483-494, ISSN 0261-5177

Tasci, A.D.A. & Denizci, B. (2010). Fashionable hospitality: A natural symbiosis for Hong Kong's tourism industry? *International Journal of Hospitality Management*, Vol.29, No.3, pp. 488-499, ISSN 0278-4319

Taylor, S. A. & Baker, T. L.(1994), An assessment of the relationship between service quality and customer satisfaction in the formation of consumers' purchase intentions. *Journal of Retailing,* Vol.70, No. 2, pp.163-178, ISSN 0022-4359

Timothy, D. J., & Butler, R. W. (1995). Cross-border shopping: A North American perspective. *Annals of Tourism Research,* Vol.22, No.1, pp. 16-34, ISSN 0160-7383

Tourism Bureau M.O.T.C., Taiwan R.O.C. (2011). *2010 Annual Survey Report on Visitors Expenditure and Trends in Taiwan.* ISBN 978-986-01-9401-2, Taipei: Tourism Bureau M.O.T.C., Taiwan R.O.C.

Xu, Y., & McGehee, N. G. (2011). Shopping behavior of Chinese tourists visiting the United States: Letting the shoppers do the talking, *Tourism Management,* doi:10.1016/j.tourman.2011.05.003.

Yalch, R. F., & Spangenberg, E. (1990). Effects of store music on shopping behaviour. *Journal of Consumer Marketing,* Vol.4, No.1, pp. 31-39, ISSN 0887-6045

Yuksel, A. (2004). Shopping experience evaluation: A case of domestic and international visitors. *Tourism Management,* Vol. 25, No. 6, pp.751-759, ISSN 0261-5177

Yuksel, A. (2007). Tourist shopping habitat: Effects on emotions, shopping value and behaviours. *Tourism Management,* Vol.28, pp.58-69, ISSN 0261-5177

9

The Case Studies of Using HR Practices for Improving SQ from Various Service Typologies

Jui-Min Li

Central Taiwan University of Science and Technology,
Taichung City, Taiwan
Republic of China

1. Introduction

In developed countries, service industry production values exceed those of their manufacturing industry. In a report on the service industry production values of various countries, released by the Taiwan Directorate-General of Budget, Accounting and Statistics in June 2011, the United States service industry production value was 73.6% of GDP in 2009. In the same year, Germany, Japan, and Taiwan were contributed to by the service industry at 75.6%, 72%, and 69.3%, respectively. These figures demonstrate the importance of the service industry to the economies of industrialized countries.

In whichever form they may be, they must transmit services to customers through service personnel. For this reason, human resources (HR) are important to the service industry, especially service typologies with high levels of contact between service personnel and customers. However, for different high-level contact services, customers have different expectations of the content and competency levels of the service personnel. For instance, patients in hospitals need professional medical care. In leisure and entertainment, customers expect the service personnel to be able to create an enjoyable and fun atmosphere, with the emphasis being put on the emotional side of service.

When it comes to quality management (QM) in the service industry, the most fundamental component is service quality (SQ), since the basis of service lies in physical equipment and personal interaction (Thomas, 1978). Thus, in order to upgrade SQ, the approach may be either through the system or through HR practices. However, in the past, studies in QM focused on the system factors, thus causing a gap on the topic of HR management (Soltani, Van, & Williams, 2004).

HR practices are a significant issue in QM. Studies have found that when employees have better skills and higher motivation, they are able to offer a better service, leading to greater customer satisfaction (Hennig-Thurau, 2004). All these point to the fact that HR practices are potent tools for improving SQ. In fact, HR practices are "the other side of quality" (Wilkinson, 2004).

Two companies and one clinic have been selected from various service typologies. This paper concerns the function of HR practices in QM, and describes how to improve the competencies of service personnel through HR practices and, in turn, advance their SQ.

2. Literature review

The service provider can improve SQ from the service system and through the competencies of service personnel (Li et al., 2008). Li et al. (2009) argued that the competency of service personnel is divided into three important elements: knowledge and technical skills, social skills and service attitude.

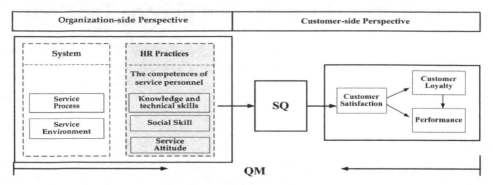

Fig. 1. Relationship between HR practices and QM

2.1 Knowledge and technical skills

These refer to the professional abilities of service personnel in providing services, i.e., the tacit knowledge they must possess to complete their tasks and solve problems (Spencer & Spencer, 1993). For instance, a surgeon must possess knowledge of human physiology and anatomy, as well as the various principles of medical therapy. Moreover, service personnel must, likewise, possess the technical skills needed for carrying out their tasks (Spencer & Spencer, 1993). For example, surgeons must be able to actually operate on patients and teachers must be able to transmit knowledge to their students. It is only after becoming acquainted with the components of hair, the features of hair products and having been equipped with the skills of hair cutting and hair dyeing that hair stylists manage to complete the task of hairstyling.

2.2 Social skills

Some services require a high level of interaction between service personnel and customers in order to determine the latter's needs, thus allowing the former to complete the service. At this time, service personnel possess the communicative abilities of listening and expressing themselves, which lead to sound interaction with customers, the understanding of the customers' needs and an explanation of ideas (Arthur & Bennett, 1995). Social skills are defined in this way: service personnel, well-oriented to their tasks, may use linguistic communicative competencies on customers to help solve individual demands, explain the service process or answer questions, and maintain a good relationship with their customers (Li et al.,2009).

2.3 Service attitude

This includes inherent work motivation (Arthur & Bennett, 1995) and enthusiasm (Spencer & Spencer, 1993). Service attitude is defined as the internal passion for a job. In the process of providing services, personnel use words of concern or facial expressions, a smile and other such non-linguistic expressions to serve customers with an affective attitude (Li et al.,2009).

In addition, Li et al. (2009) proposed that perceived risks influence the level of competencies needed by service personnel. Based on three different levels of perceived risk, this chapter selected a department store, a chain of hair-salons, and an ophthalmologist clinic that all offer excellent SQ to describe how HR practices can improve SQ, and the differences in HR practices among these industries.

3. Case I- Department store industry

The department store industry falls under the low risk service type. Case I is the second-largest department store in central Taiwan. It was once ranked No.1 for the level of customer satisfaction among all the department stores in Taiwan (opinion poll conducted by China Productivity Center in 1995). In 2001, customers assessed it as the department store with the best image in central Taiwan. Also, in 2006, it was ranked as No.3 in the 'best brand' category for department stores in Management magazine in Taiwan. It was also ranked number 3 in a service industry survey conducted by Global Views Monthly magazine in 2009.

3.1 Data collection

Data collection involved interviews, observation, document analysis, and surveys to proceed with triangulation.

The first observation method involved Consensus Camp and reading and discussion circles. Second, interviews were conducted with the trainees and organizers. Third, document analysis includes the information released by the Securities and Exchange Commission (including financial data, company regulations, and announcements), magazine reports, news, institutional reports, and academic theses related to the company. Internal data includes teaching materials, standards and recorded activities in the reading and discussion circles, Consensus Camp proposals, PowerPoint slides and records of meetings, organizational charts, and other related materials. Finally,surveys on SQ in 2005, 2006 and 2011 were adapted. The research used Retail Service Quality Scale (RSQS)(Dabholkar et al.,1996), which is designed mainly for retail. 700 questionnaires were distributed and collected at the entrance to the company in 2005 with 629 valid questionnaires returned. A further 300 questionnaires were given out and collected in 2006 with 287 valid questionnaires. In 2011, 211 questionnaires were issued and collected with 201 valid questionnaires returned. All items were measured with a 5-point Likert scale. The Cronbach's a values are 0.91 ,0.93 and 0.94 in 2005,2006 and 2011, respectively.

3.2 HR practices

The retail department store industry belongs to a type of low-risk service mode, where service providers place more emphasis on social skills and service attitude。The three practices adopted in case I are as follows.

Reading and Discussion Circle	1. Trainee : All managers 2. Period : From October, 2001 until now 3. Process: CEO and lecturers within HRM department lead trainees of every class to repeatedly read and understand texts as well as guides managers in every level to strengthen their Chinese philosophy.	
Common Consensus camp	1. Trainee : employees within the company and salesperson of factories in 2005-2006 2. Period : From March, 2003 until now 　　　Held between November and December every year 3. Process : A two-day event takes place outside the company. CEO and internal lecturers teach Chinese philosophy to all staff to enhance interaction and understanding.	
Departmental Reading and Discussion Circle	1. Trainee : Employees in every department 2. Period : From December, 2007 until now 3. Process : Employees should spread what has been learnt in Reading and Discussion Circle when working in their departments. In every department, employees can read texts and articles and watch films with others so that what each has learnt and experienced can be shared with others.	

Note:

⊕　Stands for CEO and HRM Department

　Stands for managers

○　Stands for employees

Table 1. Three Training Methods

3.2.1 Knowledge and technical skills

- Objects: Managers
- Process: shared learning
- Activity: Reading and Discussion Circle
- Location: in the company
- Frequency: Once a week

3.2.2 Social skill

- Objects: Employee, Managers, Sales of Factory
- Activity: Reading and Discussion Circle,Common Consensus Camp
- Process: Sharing personal experiences and ideas, features a host
- Location: Company headquarters, off-site
- Frequency: Once a week, Once a year.

3.2.3 Service attitude

- Objects: Employee, Managers
- Activity: Reading and Discussion Circle, Common Consensus Camp
- Process: Sharing personal experiences and ideas, features a host
- Frequency: Once a week, Once a year.

3.3 Effects of QM

The most fundamental component is SQ when it comes to QM in the service industry. The mean of SQ started from 3.69 in 2005, rose to 3.80 in 2006 and to 3.93 in 2011, which means that SQ has significantly improved with a F value of 24.71 (p value= 0.000***), as shown in Table 2. The company expected to enhance the employees' work spirituality through the training and the SQ increased. Regarding personal interaction in SQ, its improvement has surpassed all other aspects with significant growth each year. The means of personal interaction are 3.58 in 2005, 3.78 in 2006 and 3.95 in 2011 (F=24.71, p= 0.000***), indicating that spiritualization training can change employees' service attitudes and, subsequently, lead to improvements in SQ.

Year	N	SQ		Personal Interaction		Physical Aspects		Reliability		Problem Solving		Policy	
		Mean	S.D.	Mean	S.D.	Mean	S.D.	Mean	S.D.	Mean	S.D.	Mean	S.D.
2011	201	3.93	0.46	3.95	0.57	3.98	0.51	3.93	0.55	3.86	0.56	3.86	0.54
2006	287	3.80	0.44	3.78	0.53	3.88	0.51	3.85	0.51	3.71	0.59	3.75	0.56
2005	629	3.69	0.43	3.58	0.59	3.80	0.46	3.69	0.50	3.56	0.56	3.72	0.53

Table 2. The Statistical Description of Each Aspect of SQ

By featuring HR practices, Case I places the emphasis on increasing employees' flexibility in order to enhance their service attitudes and SQ rather than their knowledge and technical skills. The service attitudes of frontline employees' (FLEs) are of great importance, but it requires a long time to cultivate and is not quite as effective in the short term. Therefore, few enterprises are willing to put it into practice. Case I is endowed with the features that can increase employees' flexibility to promote their service attitudes.

4. Case II- Hair-salon industry

The hair-salon industry is of medium risk. Service personnel for the hair-salon industry must be equipped with relatively medium-skilled techniques and high socialization skills (Li et al., 2009). This case study delineates how the hair salon chain integrates HR practices, to help improve employees' competencies.

4.1 Data collection

Data were obtained through secondary data and interviews. Secondary data included the following: the founder's biography, company website, salon management manuals, journals from the HR department, handbook of hairdressing lessons, films used for internal training, magazine reports, etc. The target of the interviews was a manager who worked in the company's HR, administration and business expansion departments. The author's familiarity with this executive not only made the acquisition of data and their verification easier, but also led to greater accuracy in the information obtained from the interviews. Seven interviews were conducted over a period of eight months, each one tape recorded and transcribed verbatim into written versions.

4.2 HR practices

Before we discuss the HR practices adopted, it is necessary to first study how the company classified their trainees. They are classified into three levels: (1) Apprenticeship; all hair designers have to start at this level in which they work on service steps requiring lower techniques. (2) Hair designers, who handle the core service process, and (3) Managers, who take charge of the daily operation of the salons. HR practices for these three types of trainees are shown in Figure 2.

4.2.1 Knowledge and technical skills

Four training approaches are used for the three types of service personnel, as follows:

1.   Apprentices

1.   Apprenticeship
•   Content: Technical skills.
•   Trainers: A multi-mentor approach is adopted through which apprentices take turns to learn from various hair designers.
•   Process: Observation, imitation and learning by doing.
2.   Programmed instruction:

Content: Basic-technique teaching materials are prepared by the HR department at the company headquarters, covering professional knowledge and actual techniques.

Trainers: Internal instructors in all salons and from the company headquarters.

Process: Each level is assigned a fixed number of hours and adopts the "learning passport" approach that allows the completion of the prescribed training hours in different locations.

2.   Hair designers

1.   Theme Training:
•   Content: Transmission of latest trends, new techniques and marketing concepts, etc.
•   Trainers: Experts from outside the organization. Foreign techniques are adopted to maintain market leader legitimacy.
•   Process: Newly promoted hair designers are required to complete a "wandering leave" in which they work in other salons for three months. Training is done in three ways. First, external instructors give lessons on the latest trends and customization. Second, instructors assigned by suppliers teach about new products and techniques. Third, overseas study tours are organized for attendance at exhibitions and competitions held in Japan, Europe and the US, as well as engage in technique exchange.

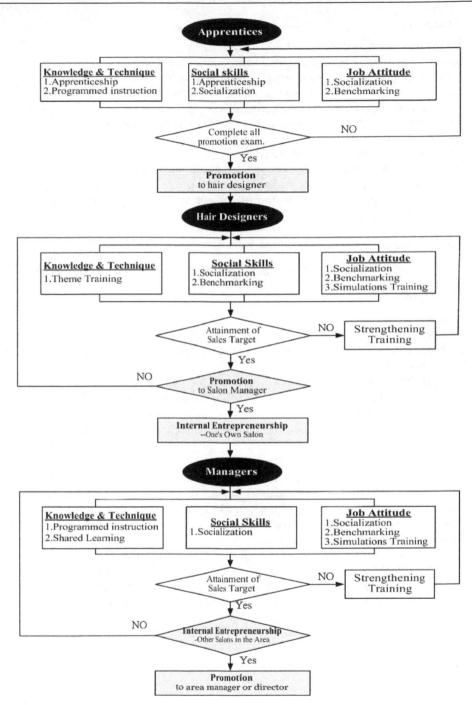

Fig. 2. The major content of HR practices

3.   Managers

1.   Programmed Instruction:
*   Content: Salon operation issues and management concepts.
*   Trainers: Documents, senior managers, external instructors or consultants.
*   Process: For salon operation management, operating procedures are standardized and documented. For business management, concepts on leadership, motivation, team cohesiveness and marketing are taught in province-wide managers' meetings.
2.   Shared Learning:
*   Content: Actual operational experiences.
*   Process: Periodic managers' meetings to discuss salon operation issues, sharing of experiences in operations techniques for use on customers and employees. Area managers and directors discuss the orientation of future operations and engage in exchanges.

4.2.2 Social skills

Five training approaches are used for the three types of service personnel, as follows:

1.   Apprentices

1.   Apprenticeship:
*   Content: Technical and social skills.
*   Trainers: A multi-mentor approach is adopted through which apprentices take turns to learn from various hair designers.
*   Process: Observation, imitation and learning by doing.
2.   Socialization: The goal of this method is to influence employee behavior through interaction between each other, such as singing, slogan chanting, and calisthenics at the start of the day, or the wearing of uniforms for creating a common language.

2.   Hair designers

1.   Socialization: The goal of this method is to influence employee behavior through interaction between each other, such as singing, slogan chanting, and calisthenics at the start of the day, or the wearing of uniforms for creating a common language.
2.   Benchmarking: The goal of benchmarking is for the trainer to provide models for crucial behavior to be imitated by trainers. Senior personnel with excellent sales performance share their motivational ideas with hair designers, as well as their sense of identity with the organization, which serve as targets for imitation. For salon managers and area managers, excellent-sales performance managers are invited to talk about their service attitudes and experiences during meetings. Well-known public personalities are also invited to share their secrets for success.
3.   Strengthening Training: Hair designers who fail to meet basic quotas receive extra training to help them improve performance. Lessons on marketing and communication skills, latest trends and service attitude are taught by internal instructors.

3.   Managers

Training activities to improve service attitude are done in three ways, as follows:

1. Socialization: The goal of this method is to influence employee behavior through interaction between each other, such as singing, slogan chanting, and calisthenics at the start of the day, or the wearing of uniforms for creating a common language.
2. Strengthening Training: For managers of salons with poor sales performance, the reasons are subjected to analysis and the proper recommendations are made. The goal is to improve their sales performance.
- Content: Leadership, customer relations, analysis of salon operation methods, etc.
- Trainers: Managers with high sales performance.
- Process: Holding meetings to analyze the reasons for failure. After trainees complete their reports, all managers participate in the analysis and make concrete recommendations.

4.2.3 Service attitude

The contents of the service attitude training can be classified as policy level and training activities. The concrete ways of implementation in the company are explained as follows:

1. Policy Level

Regarding policies, service attitude is improved in two ways: promotion and internal entrepreneurship.

1. Promotion: Promotion is adopted to solve work boredom brought about by specialization and standardization. It is carried out in the following ways: (1)More promotions in the more boring jobs. (2)The promotion of technical personnel follows a fair and clear evaluation system. (3) Promotion is fused with the reward system. (4)Promotion proceeds this way: apprentice → hair designer → manager.
2. Internal entrepreneurship: Internal entrepreneurship is an approach in which employees invest money to join the salon chain and manage a salon themselves. The goal is to retain personnel with great sales performance abilities while at the same time solving managerial shortage problems and basic sales volume issues. This approach is targeted towards managers only.

2. Training activities

Training activities to improve service attitude are undertaken in three ways, as follows:

1. Socialization: The goal of this method is to influence employee behavior through interaction between each other, such that it is oriented towards company rules, procedures or culture. Through informal rites, personnel motivation is enhanced.
2. Benchmarking: The goal of benchmarking is for the trainer to provide models for crucial behavior to be imitated by trainers. Senior personnel with excellent sales performance share their motivational thoughts with hair designers. For salon managers and area managers, excellent-sales performance managers are invited to talk about their service attitudes and experiences during meetings.
3. Simulation training: The goal in this approach is to strengthen personnel motivation and work values. This involves attending classes for realizing potential through inspiration sharing and group fun activities with the goals of stimulating positive feelings, motivation, the need for self-improvement, dynamism, setback management, etc.

4.3 Effects of QM

Case II has more than 350 branches employing more than 3,000 people. Every year, they have an annual sales income of more than US$107 million derived from nearly 4 million customer visits. The company enjoys the largest market share in Taiwan's hair-salon industry. The sample company was awarded with *Taiwan Top Brands in Service and Trade* award in 2010, issued by the Division of Commerce, Ministry of Economic Affairs. After analysis, in terms of market share and awards, the subject has a better SQ as far as customers are concerned.

5. Case III- medical care industry

The third case looked at in this study was the medical care industry, which, compared to the retail department store or the hair-salon industry, is of high risk.

5.1 Data collection

The data was mainly acquired through interview and document analysis; the former was achieved by interviewing the Dean of an ophthalmology clinic in Taiwan. He was also the chief director of the Teaching and Research department at a teaching hospital, executive director of a relevant regional union, (being responsible for the review of physicians at the Bureau of National Health Insurance), and was familiar with clinical medicine and the medical environment in Taiwan. A total of three interviews were conducted, each lasting 90 minutes.

Sources of data acquired through the research method of document analysis included: National Taiwan University School of Medicine, Taiwan Joint Commission on Hospital Accreditation, and the Academic Report of Medical Treatment, etc.

5.2 The cultivation process of doctors

For the applicants who expect to provide services in medical treatment, doctors in particular, they need, at the very least, to be educated to bachelor degree level, as issued by national government. The path to becoming a doctor shall be covered from students majoring in medicine, clerk, intern, Resident(R), Chief resident (CR), Fellow (F), and visiting staff (VS)(see figure 3). The following section describes the process of cultivation for doctors.

5.2.1 Knowledge and technical skills

1.   Medical students

At this stage, the main goal for students is to study the fundamental theories in medicine. In Taiwan, a major in medicine is very popular with applicants and top students can be recruited into a department of medical science. Basic medical knowledge is passed in the first four years.

*   Content: Basics in medicine, such as Pathology, Epidemiology, and Pharmacology.
*   Process: Programmed instructions are adopted. The courses at this stage are well designed by course committees and are also ratified by the administrative office of the department and the university. In addition, these curricula have to be thoroughly examined by the Evaluation Committee of the Ministry of Education every three years.

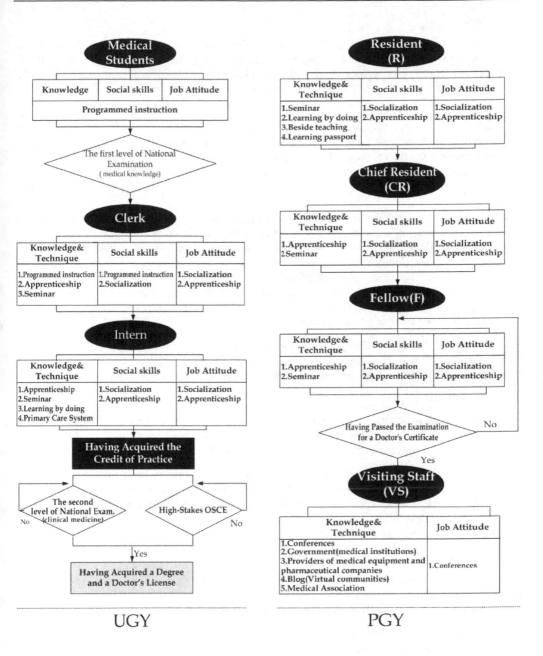

Fig. 3. Process of cultivation for doctors
Note: OSCE stands for Objective Structured Clinical Examination
UGY stands for Under-Graduated Year
PGY stands for Post-Graduate Year

2.   Clerk

After having passed the National Basic Medical Subjects Examination, students majoring in medicine can join the clinical training of a Clerk. During this time, fifth and sixth grade medical students not only receive clinical knowledge in school, but also attend a clerkship training program of internal medicine, surgery, ambulatory medicine and pediatrics in teaching hospitals.

- Content: Inquiry, physical examination, logical thinking and basic clinical practice training, such as treatment decision-making, and the concept of clinical skills (diagnosis, treatment procedures, medical writing, aseptic technique, scrubbing, dressing).
- Process: Apprenticeship, programmed instruction, seminar.

3.   Intern

After about 1 or 2 years of clerk training, sixth grade or seventh grade students majoring in medicine then become Interns. They attend clerkships in different divisions of the hospital to acquire practical clinical skills through the guidance of physicians. Although they have not yet acquired a doctor's license, they can do some simple medical practices as long as they have the presence of guiding physicians. These practices may include inquiry and examination after the admission of patients, injections, medication, intravenous fluids, inserting a nasogastric tube, inserting catheter, and the suture of a small wound, etc. This training will last a further 1 or 2 years.

- Content: Practice of clinical techniques (implementation of diagnostic procedures, treatment procedures, medical writing, aseptic technique, scrubbing, dressing)
- Process: Apprenticeship, seminar, learning by doing, Primary Care System

4.   Resident (R1-R3)

Resident Physicians are professional trainees in the hospital who have already acquired a Bachelor's Degree and have passed the physician licensing examination. A Resident Physician in the first year is called R1, the second year, R2, and the third year, R3.

- Content: PGY basic training courses, practical training in general medicine, physical examination, community medicine (including outpatient, emergency or fever screening stations), basic medicine (internal medicine, surgery), and relevant elective courses and holistic medical training.
- Process: Seminar, learning by doing, bedside teaching, learning passport.

5.   Chief resident (CR)

A CR position is usually taken by a fourth year senior resident (R4) who is responsible for the administration of the Branch, arranging residency, internship programs and work-shift rotas, and chairing the teaching-purpose morning conference etc.

- Content: Clinical medicine
- Process: Apprenticeship, seminar.

6.   Fellow(F)

This is the second phase of resident physician. After having received three years of professional training in the traditional four divisions (internal medicine, surgery, gynecology and pediatrics) and having obtained a specialist's license, Fellows must receive further training as sub-specialists and at this stage they are called research physician. The majority of their work at this stage is to study the sub-specialist expertise and techniques. In addition, they must also be responsible for sub-specialty consultation and clinic teaching. Any patient who needs a consultation in this sub-specialty will usually be first visited by a Fellow. After this training phase, the Fellow can take the sub-specialist license exam.

- Content: Sub-specialty knowledge and technique, sub-specialty clinical practice
- Process: Apprenticeship, seminar.

5.2.2 Social skills

A physician's social skills include ascertaining the patient's symptoms, explaining the state of an illness, pacifying patients' emotions and building trust between doctors and patients.

1. Medical Students

- Content: Communication and Expression, Teamwork
- Process: Programmed instruction

2. Clerk

In the phase of Clerk, the studying sites of clerks cover both universities and hospitals. In addition consistent training of communication and expression skills is undertaken with students learning how to interact with patients and team mates in hospitals.

- Content: Communication and Expression, Teamwork
- Process: Programmed instruction, socialization
- Location: University or Hospital

3. Intern

At this stage, most of the time will be spent in hospital practice. Interns learn practical social skills and the skills to write medical records from the mentoring relationship and the process of socialization.

- Content: Communication and Expression
- Process: Apprenticeship, socialization
- Location: Universities or Hospitals

4. Resident

With the guidance of mentoring physicians, Resident physicians learn how to interact with patients, establish doctor-patient relationships, inquire about the medical history of patients, explain the state of an illness, and independently write medical records.

- Content: Communication, listening attentively and expression (including literal expression); establishing good relationships between doctors and patients.
- Process: Socialization, apprenticeship

5.   Chief Resident

A CR is mainly responsible for the teaching programs and work-shift rotas of Resident and interns together with the chairing of seminars.

- Content: Interpersonal interaction, expression competencies
- Process: Socialization, apprenticeship

6.   Fellow

- Content: Expression abilities
- Process: Socialization, apprenticeship

5.2.3 Service attitude

The cultivation of job attitude is mainly about focusing on the issues of professionalism and ethics.

1.   Students in Medicine

- Content: Clinical Ethics
- Process: Programmed instruction

2.   Clerk

- Content: Clinical Ethics and Law
- Process: Programmed instruction; apprenticeship

3.   Intern

- Content: Clinical Ethics and Law
- Process: Apprenticeship, socialization

4.   Resident

- Content: Clinical Ethics and Law
- Process: Apprenticeship

The HR Practices of Case III

After having become VS, a physician shall be equipped with the knowledge and technical skills that will include: specialist professional knowledge and new techniques, medical regulations and national health policy, national health insurance payment norms. A license renewal requires attendance at 180 hours of classes over a period of six years, of which 90 per cent are on specialized medical courses. Therefore, this case mainly introduces new professional knowledge. The methods adopted in Case III are acquired through medical seminars, the introduction of new products from medical instrument companies or drug manufacturers, medical associations and blogs, etc.

Social skills refer to patient relationship management. Case III is a typical one in which people learn from each other in a social network that includes blogs, communities, and Medical Associations.

Service attitude refers to the physicians' work motivation towards their medical career and their enthusiasm with patients. Case III indicates that after the accomplishment of physician training courses, the relevance between a doctor's personality and their service attitude is relatively high after independent medical practice.

Here we should note that the above mentioned blog is a online community where doctors in Tawan can exchange messages. The content of messages exchanged includes medical regulations and policies, the review system for national health insurance, medical malpractice, discussion of medical news, tax information, education information, introduction of deceased physicians' biographies, and mutual encouragement (for example, a clinic a clinic burns down or express their support to event figures) etc. Blog members are classified into different levels based on their anticipation level in this online community, ranging from medical students to VS. In addition, in the regional Medical Association, those physicians who are enthusiastic and have a good performance are more likely to be elected as leaders. Case III actively interacts with the virtual blog in which he actually plays the role of a VS and has also been elected as an executive member in the regional union.

5.3 Effects of QM

The respondents include a large number of patients. Based on the data provided by the Bureau of Health and Department of Health in Taiwan, each doctor in Taiwan has a caseload of 546 patients in a typical month. Every month, the ophthalmologist attends to 2500-3200 patients in this clinic.

This clinic was rated as a top clinic by the District Public Health Department. The Public Health Department occasionally invites university professors and health officials to evaluate the clinics in this area. The evaluation results for Case III are excellent. Therefore, from the number of patients and the government evaluation results, the clinic (the ophthalmologist) offers a high standard of medical treatment.

6. Discussion

This paper, from an organizational perspective, depicts three different types of HR practices undertaken by three different services. The following sections discuss further the similarities and differences between the present study and the study of Li et al. (2009) and the sources of the differences.

6.1 Knowledge and technical skills

Li et al. (2009) pointed out that among the three types of services, the knowledge and technical skills of FLEs in the medical industry are the most valued by customers. This fact is consistent with the study result of this paper. Why? From the perspective of organizations, it is the complexity of the tasks. However, from the customers' point of view, it is the perceived risk to the customer. Medical practices are the most complicated among the three industries: Physicians must acquire the basic concepts of structure and functions of the human body, pathology, pharmacology, and also receive long-term practical training in order to provide medical services. Case III shows that in order to acquire a doctor's certificate, the applicants shall pass a written test in basic medicine and clinical medicine held by the national government, together with six to seven years of clinical and sub-specialty training, before being able to conduct medical practice.

From the customer's perspective, department store FLEs cannot explain clearly the product features or components, meaning that this type of service has the least damage to the money paid by customers or to their body. During hair cutting or hair-dyeing, hair stylists can redesign or expand the service time if they cannot satisfy their customers' demands. On the other hand, if a doctor did not give a correct diagnosis (such as failing to check a cancer) or fail to offer appropriate treatment (taking surgery as an example), the consequences could endanger the patient's life. Therefore, customers hold that the perceived risk in the medical industry is the highest among the above-mentioned three industries.

Among the three cases, when imparting knowledge, the medical industry is the most systematic, complete and has the longest history. Clinical technical skills training is the longest and most solid. As a result, the knowledge and technical skills competency of the medical staff is of great importance.

6.2 Social skills

Methods of curriculum planning, learning by doing and apprenticeship can be utilized to enhance social skills. Comparatively speaking, learning by doing and apprenticeship are much more useful and practical. The more complex the tasks, the more suitable the apprenticeship is. That is to say, learning by doing is much more appropriate.

Through social skills, FLEs in the department store industry can determine what customers want or need and thus introduce the right products to customers and describe product features. Case I can train and promote an employees' social skills of listening attentively and expression using the method of learning by doing.

Hairdressing services require hairstylists to make a reasonable judgment of the most appropriate hairstyle according to customers' verbal and physical behavior and after a full consultation with their customers. Therefore, their task complexity is at a moderate level. Case II, through apprenticeship, allows apprentices to observe the hairstylists for a long time and learn how to design and be flexible in line with the preferred hairstyles of their customers.

The medical industry is the highest in terms of task complexity. Physicians are expected to make a reasonable judgment of the possible illness of patients by observing their symptoms or their statements and make further enquiries in order to make a treatment decision. At the same time, physicians are required to interpret the disease and to remind patients. Only in this way can physicians establish a reliable doctor-patient relationship. Such an interactive process needs long-term support from the cultivation system of apprenticeship.

6.3 Service attitude

Service attitude is defined as an internal passion for a job. In the process of providing services, the FLE uses words of concern or facial expressions, a smile and other such non-linguistic expressions to serve customers with an affective attitude (Li et al.,2009).The service attitude of FLEs in services sector is therefore of great importance to SQ, although it is hard to improve. The present paper mentioned three processes in promoting service attitude, including programmed instruction, apprenticeship and socialization. Programmed instruction occurs only in university education, such as a university curriculum for medical students. Apprenticeship is employed for industries that require skillful techniques, such as hairdressing and medical industries. Besides learning the techniques from a more experienced tutor, apprentices also learn their service attitudes from their mentors. Socialization is applicable to all industries.

From the prospective of organizations, hospitals or clinics, the lower the level of risk (the lower the complexity of the task), the more emphasis shall be made on enhancing service attitude, such as in case I of the department store industry. On the other hand, the higher the risk, the less attention given to service attitude, such as physicians. That is probably because physicians are self-driven and do not need training to enhance their service attitude.

7. Conclusions and managerial implications

The conclusions are summarized as follows: the higher the customer risk (the higher the task complexity), the more emphasis is placed on knowledge and technical skills. With regards to the aspect of enhancing social skills; the more complex the task, the more suitable it is for adaptation to an apprenticeship system. On the other hand, learning by doing is much more appropriate. Apprenticeship or socialization are mostly used to enhance service attitude, the former is suitable for knowledge or technical skills and the latter one for common work.

Theoretical implications include contributions to QM and HRM theory. In contributing to QM theory, this paper describes how service personnel from various service industries influence SQ. Furthermore, this paper explains how HR practices can improve a company's QM. In addition, HRM theoretical implications include providing a new perspective by comparing three different case studies and illustrating the differences in training design and performance evaluation among service types. Our illustration of the different competencies that service personnel in different service industries must possess highlights the differences and importance of service delivery in the area of service management. Finally, changing employees' attitude is a long-term task and the result is not always obvious (Li et al., 2009). Therefore, fewer companies are willing to conduct this form of training. Case I describes how employees workplace spirituality can be promoted, which can also improve service attitude and encourage SQ and Case I demonstrates an empirical study of accumulated workplace spirituality.

Practical implications for this study include providing best practices for different types of service industries. HRM managers could select similar industries to act as references for them. For practitioners of organizational development, the content and processes of the different industries provided by this study act as references for them in the areas of training design, quality management, and organizational development.

8. References

Arthur, W. J., & Bennett, W. J. 1995. The international assignee: The relative importance of factors perceived to contribute to success. *Personnel Psychology*, 48: 99–114.

Dabholkar, P. A., Thorpe, D. I., & Rentz, J. O. 1996. A Measure of Service Quality for Retail Stores: Scale Development and Validation. *Journal of the Academy of Marketing Science*, 24(1): 3-26.

Hennig-Thurau, T. 2004. Customer orientation of service employees: Its impact on customer satisfaction, commitment, and retention. *International Journal of Service Industry Management*, 15(5): 460-478.

Li, Jui-Min, Yang, Jen-Shou, & Wu, Hsin-Hsi. 2008. Improving service quality and organisation performance through human resource practices. A case study. *Total Quality Management & Business Excellence*, 19(9-10): 969–985.

Li, Jui-Min., Yang, Jen-Shou., & Wu, Hsin-Hsi. 2009. Analysis of Competency Differences among Frontline Employees from Various Service Typologies: Integrated the Perspectives of the Organisation and Customers. *The Service Industries Journal*, 29(8).

Soltani, E., Van Der Meer, R. B., & Williams, T. M. 2004. Challenges Posed to Performance Management by TQM Gurus: Contributions of Individual Employees Versus Systems-Level Features. *Total Quality Management & Business Excellence*, 15(8): 1069-1091.

Spencer, J., & Spencer, S. M. 1993. *Competence At Work: Models for Superior Performance*. New York: Wiley.

Thomas, D. R. 1978. Strategy is different in service businesses. *Harvard Business Review*, 56(4): 158-165.

Wilkinson, A. 2004. Quality and the Human Factor. *Total Quality Management & Business Excellence*, 15(8): 1019-1024.

Section 3

Total Quality Practices to Quality Management

10

Control Systems for Quality Management

Zulnaidi Yaacob
Universiti Sains Malaysia
Malaysia

1. Introduction

Total Quality Management (TQM) has become a strategic effective weapon for the successful of various type of organizations. Given its strategic importance, TQM has been implemented in various organizations such as manufacturing (Arawati, 2005; Sohal & Terziovski, 2000; Zakaria, 1999); small medium enterprises (Mohd Nizam & Tannock, 2005); higher education (Cruickshank, 2003) as well as public departments (Abdul Karim, 1999; Hunt, 1995; Nor Hazilah, 2004). The quality management as a discipline has developed through several phases, starting from 'quality by inspection', 'Statistical Quality Control (SQC)', 'Quality Assurance (QA)', and 'Total Quality Management (TQM)' (Prybutok & Ramasesh, 2005). According to Kanji (2002), the ultimate goal of TQM is customer satisfaction. To achieve this, the implementation of TQM requires all members of an organization work as a team through the culture of continuous improvement. In other words, the three pillars of TQM are employee empowerment, continuous improvement and customer focus.

In their studies, these authors (Arawati, 2005; Li, Yasin, Alavi, Kunt, & Zimmerer, 2004) have revealed that there is a significant relationship between TQM and performance. However, not all TQM implementers have executed their TQM successfully. These contradicts findings were reported by other researchers (Samson & Terziovski, 1999; Sanchez-Rodriguez & Martinez-Lorente, 2004; Sohal & Terziovski, 2000; Witcher, 1994). In a review paper, Sila and Ebrahimpour (2002) concluded that previous studies on the relationship between TQM and performance had revealed inconsistencies and sometimes produced conflicting results. Ehigie and McAndrew (2005) concluded that the implication of these unresolved issues is that, future researchers need to investigate factors that could influence the results of TQM. Thus, more empirical evidence is needed to shed light on this unresolved issue.

The unsuccessful of strategy implementation by organizations worldwide are common stories. According to Kaplan and Norton (2000) 70 to 90 percent of organizations worldwide failed to execute their strategy successfully. Contingency theory literature suggests the unsuccessful implementation of a strategy is due to the 'mismatch' or 'misfit' between the strategy and control systems. A strategy is only a means toward an end. To ensure a strategy is successfully executed, strategy related control systems must be institutionalized (Zakaria, 1999). As widely discussed in the literature, good performance effect of TQM can be harvested by complimenting it with effective and supportive control systems (Andersen, Lawrie & Savic, 2004; Daniel & Reitsperger, 1991; Selto, Renner & Young, 1995)

In line with the contingency theory, management accounting literature proposes that the periodicity of traditional accounting control systems such as budgetary control systems has been blamed for ignoring organizational long-term initiatives (Hayes & Abernethy, 1980; Kaplan, 1983; Otley; 1999; Rangone, 1997), such as in meeting the urgency of TQM in an organization. Dent (1990) in his review paper, stressed that shortermism of traditional accounting control systems had been criticized as discouraging the long-term focus of TQM to be in place. Consequently, several contemporary complementary approaches such as strategic management accounting, strategic cost management, and non-financial performance measurement have been proposed and introduced as means to overcome the limitations of traditional management accounting control systems in dealing with strategic issues (Otley, 1999; Rangone, 1997). In addition to these approaches, the role of strategic control systems (SCS) has been recognized as important systems for the purpose of strategy implementation (Chenhall, 2005; Hoque, 2004; Muralidharan, 2004) including TQM. Hoque (2004) for instance, revealed that there is a significant association between organizational strategy and performance through the presence of SCS.

The increase in TQM awareness has been recognized as one of the important factors that is heightening interest among researchers in control systems issues (Butler, Letza & Neale, 1997). However, the existing literature on the relationship between organizational strategy and control systems is still at the beginning, incomplete and at its infancy (Chenhall, 2003; Daniel & Reitsperger; 1991; Otley, Broadbent & Berry, 1995). They shared a similar view that the issues of the relationship between organizational strategy and organizational control systems were not well addressed in the previous scholarly literature. Therefore, this research area deserves more empirical evidence in examining the relationship between TQM and control systems.

This study contributes to the academia and practitioners. Without doubt, this study is vital due to the inconsistencies, unresolved and even contradictory findings in identifying the relationship between organizational strategy, structure and performance (Prajogo & Sohal, 2006). Given this phenomenon, this study aims to provide evidence to support the proposition that the structural relationship between TQM, SCS and organizational performance is significant.

2. Problem statement

The Malaysian government has initiated various quality management strategies initiatives such as TQM, Zero Defect and Customer Charter to improve the performance of public service. The starting point of current quality awareness in the Malaysian local government was the launch of the 'Excellent Work Culture' campaign in 1989. All these quality management initiatives would be associate with various benefits including better product quality (Ahire & Golhar, 1996), shortening service delivery times, increasing customer satisfaction, as well as achieving higher productivity (Sila & Ebrahimpour, 2002).

However, after more than 20 years of the 'Excellence Work Culture' programme has been launched, the performance of local governments in Malaysia still receives much criticism and complaints, suggesting the inability of these institutions in delivering high quality services to meet the expectation of public at large (Ibrahim & Abd Karim, 2004). As such, the Minister of Housing and Local Government has made a statement that the Malaysian local government was unable to deliver a good service to the public (The Star, p.14, 6 June 2004).

Given the phenomenon, the relationship between TQM, strategic control systems and performance of Malaysian local government provides an opportunity for a scientific investigation to be undertaken. The low performance of TQM organizations could be related to the issue of misfit between TQM and control systems (e.g. Ittner & Larcker, 1997). Although the number of studies on the relationship between TQM and performance that are reported in the literature are encouraging, the issue of the inter relationship between TQM, control systems and performance has not been fully explored. Given the shortcomings in the literature concerning the issue of control systems in explaining the relationship between TQM and performance as well as emerging issue of performance of Malaysian public service, this study undertakes to investigate:

To what extent TQM is related to strategic control systems in order to achieve good organizational performance.

3. Research question and objective

In line with the research problem discussed in the preceding section, the following research question was investigated: Is the structural model of TQM and organizational performance mediated by strategic control systems? The objective of this study is to test the structural relationship between TQM, strategic control systems and organizational performance strategic control systems.

4. Literature review

4.1 Malaysian local government's experience

The number of government servants in Malaysia is increasing year by year, with the current number of more than 1.3 million servants as depicted in Table 1. Apart from the high number of employees, the public sector has also contributed a significant rate of Gross Domestic Product (GDP). As such in 1997, public sector contributed 6.6 % out of the total Malaysian GDP, while in 2002 the percentage of contribution into GDP by the public sector increased to 7.2 %. Given the high number of total employment and percentage of contribution to GDP, it is evident that the public sector forms an enormous part of the allocation of Malaysian nation wealth. Moreover, this sector is also accountable for its action to every citizen of the nation. Thus, the importance of having effective public organizations to be in place is clearly apparent.

In Malaysia, we have a three-tier system of government which comprises of the federal government, state government and local government. Local government falls under the exclusive jurisdiction of the respective state government. Evolving from a major restructuring carried out in 1973, local governments today come within two principal categories, which are the Municipal Council and the District Council (Abdul Karim, 1999). But nowadays, several local governments have achieved the status of City Council. Based on the information provided by the Ministry of Housing and Local Government, there are 145 local governments in Malaysia as in December 2005 (http://www.kpkt.gov.my).

Date	Number of Government Servant
31 December 2000	979,464
30 June 2001	985,967
31 December 2001	994,548
30 June 2002	1,004,508
31 December 2002	1,026,143
30 June 2003	1,060,649
31 December 2003	1,080,886
30 June 2004	1,041,778
31 December 2004	1,098,638
30 June 2005	1,337,413

Source: Official website of Public Service Department, http: www.jpa.gov.my/statistik (accessed on 26 July 2006).

Table 1. Numbers of Government Servant by Year

As a forefront organization, performance of the Malaysian local government becomes very crucial (Ibrahim & Abdul Karim, 2004). As with other public organizations, local government in Malaysia also response to any new improvement agenda promoted by the federal government. One of the significant improvement agenda was the birth of the quality initiatives. The impetus of current quality awareness in our public sector is the launch of the 'Excellent Work Culture" campaign in 1989. Since then, the Malaysian public sector has embarked on various administrative reforms like TQM, zero defect and ISO 9000. In general, the commitment of public organizations to the implementation of quality initiatives has been very encouraging (Muhamad et al., 2003).

The launching of the 'Excellent Work Culture Movement' by the Rt. Hon. Prime Minister on 27th of November 1989 showed the Malaysian government's commitment towards quality and productivity improvement in the public services. Subsequently, under Excellent Work Culture Movement, the government has instituted various activities for improving quality in the public sector. Among the quality initiatives was the launching of the Manual on Quality Management and Improvement in the Public Service on 25 June 1990 (DAC No 4 of 1991). Further, in order to assist Heads of Department in implementing strategies for quality improvement in their respective agencies, the DAC No 4 of 1991, entitled 'Guidelines on Strategies for Quality Improvement in the Public Service' was published. As elaborated in this guideline, government agencies are given the responsibility for planning and implementing seven programmes for quality improvement, namely Q suggestion system, Q process system, Q inspection system, Q slogan, Q day, Q feedback system, and Q information system (DAC No. 4 of 1991).

As with other Malaysian governmental organizations, many local governments in Malaysia over the past years have committed to institutionalize TQM. As commonly reported in TQM literature, a number of benefits can be derived from the implementation of TQM into the local government. For instance, the institutionalization of TQM shows to the community at large that the local government is committed to improve the performance. As prescribed by

TQM scholars, TQM is among others, able to ensure that local government is responsive to the expectations and needs of its constituents; empower its servants in performing their responsibilities; create a conducive working environment; and achieve the objective of cost effective (Hunt, 1995).

However, the implementation of TQM in the public organizations is not free from several obstacles due to the nature of public service. Generally, customers of public organizations are various, key performance indicators are problematic to be measured precisely, annual budget circles and the need for the politicians to show short-term results often devastate the long-term perspectives of TQM, and implementing TQM is time-consuming. Additionally, TQM implementation can be very expensive to the public expenses (Hunt, 1995). However, the primary obstacles for any TQM implementation are short-term vision of leaders and their lack of understanding of TQM (Sohal & Terziovski, 2000). These difficulties are reflected by the low performance of TQM initiatives in certain Malaysian local governments. As such, only two local governments, which are the City Hall Kuala Lumpur and the Kuantan Municipal Council, were nominated for PMQA (Prime Minister Quality Award) in 1990 (Government of Malaysia, 1991).

In mobilizing quality initiatives in public sector, eleven various projects have been implemented (Government of Malaysia, 1991) such as:

a. The publication of the 'Manual on Quality Management and Improvement in The Public Sector';
b. The distribution of the circular letter from the Chief Secretary to the Government (P.M 17479/11 Vol. 2 dated 30th July, 1990);
c. Training on Quality Management by Institut Tadbiran Awam Negara;
d. Production of the Training Manual on Quality Management and Improvement;
e. Talks on Quality Management by Malaysian Administrative Modernisation and Management Planning Unit (MAMPU);
f. The production of videos tapes on quality and productivity management;
g. The publication of journal of 'KHIDMAT' and 'CEKAP';
h. The Prime Minister's Quality Award for the Public Sector; and
i. Quality Control Circles (QCC).

The 'Manual on Quality Management and Improvement in The Public Sector' highlights various important concepts of quality management for government agencies. These concepts are reference points for implementing quality management and improvement efforts, the structures that need to be set-up and the way to operationalise them in the actual working environment. These concepts are (Government of Malaysia, 1991):

a. Quality is defined as meeting customers' and stakeholders' requirements which can in turn be translated into standards of excellence;
b. Quality is achieved through prevention. It can be attained by setting standards and ensuring that these standards are adhered to;
c. The performance standard is zero defect. In other words, the standard set must be achieved the first time and every time;
d. The cost of quality is the extra cost incurred due to non-conformity to standards. For instance, repetitive jobs, scraps, compensation payments and managing complaints; and

e.   All work is process. These processes can be broken down into main activities which can be illustrated through work flow charts.

Furthermore, DAC No. 1 of 1992 suggests that Malaysian Government departments must emphasize on seven management principles of TQM, namely: support of top management, implementation of long-term strategic plan on quality, customer focus, providing training and recognition, fostering teamwork, establishing performance measurement and emphasizing quality assurance. These seven management principles are in line with the critical factors of TQM as discussed in TQM literature that were scientifically developed.

4.2 The calls for strategic control systems

According to Juran (1988), the implementation of quality programme in an organization can be phased into three main phases, namely quality planning, quality control, and quality improvement. During the quality planning stage, an organization prepares to meet the intended quality objectives. Then, quality control is designed to ensure that the quality objectives set in the planning stage are being achieved at the end of the production process. The third phase of trilogy, known both as quality improvements and Juran's *breakthrough sequence*, is the means for managers to find and remedy the basic cause leading to a quality failure. In other words, the breakthrough process is used as a troubleshooting tool to keep the quality planning-control sequence running as intended.

However, Juran (1988) estimates that approximately 80% of the problems identified with breakthrough analysis are correctable only by improving management control systems. The remaining 20% can be attributed to workforce error. In other words, the existence of suitable management control systems is vital for organizations to be able to successfully implement TQM strategy. Perhaps, this situation could be more critical in public organizations, since 95% of errors are caused by systems error and only 15% are attributable to the actions of the workers (Koehler & Pankowski, 1996).

The role of third part of Juran's (1988) trilogy is to find and cure the basic cause leading to a quality failure. Conversely, the function of management control systems is to prevent those bad causes from happening (Merchant, 1982). Since the focus of TQM is prevention, therefore, if public organizations improve its control systems, they can prevent mistakes in the systems (Koehler & Pankowski, 1996). Thus, the existence of suitable control systems can be a significant system in supporting the implementation of TQM. Indeed, prevention is better than cure. However, the empirical study that examines the relationship between TQM and control systems relatively remains less explored.

Apart from many success of TQM, many researchers reported that not all TQM adopters had gained positive impact from TQM (Madu *et al.*, 1995; Powell, 1995; Yasin *et al.*, 2004). Perhaps, these failure stories were due to the critical factors of TQM considered were incompatible with the traditional mechanistic way of organizing (Hoogervorst, Koopman & Van der Flier, 2005). As commonly stressed in TQM literature, among the critical factors of TQM is teamwork and strong focus on employee involvement. As such, Juran (1995) noted that quality award winners practiced employee involvement culture to an unprecedented degree. Seeing employee as an important factor of TQM implies an approach fundamentally different from the traditional mechanistic approach. Therefore, TQM is considered to be a misfit with the traditional mechanistic organization, since it ignores employees as a crucial source for achieving strategic objectives. As elaborated by Dent (1990), mechanistic approach like traditional accounting control systems would hinder the achievement of a long-term strategy.

Conversely, Spencer (1994) advocated that many critical factors of TQM like customer focus; teamwork; management commitment; and employee empowerment were compatible with the organic organizations. Thus, the role of strategic control systems has been recognized as a vital system in supporting the organizational strategy like TQM (Goold & Quinn, 1993; Muralidharan, 2004). As documented by Goold and Quinn (1993), without strategic control systems, even good strategies can easily be blown. However, the study on the relationship between TQM and strategic control systems is relatively a new area to be explored. Thus, strategic control systems (SCS) in TQM organizations is the main focus of this study.

The organic type of organizations refers to the organizations that are implementing organic forms of control systems. Organic control systems are more flexible, responsive, involve fewer rules and standardized procedures and tend to be richer in data than mechanistic control systems (Chenhall, 2003). All of these characteristics are important in achieving customer satisfaction as the nature of customer satisfaction is dynamic. As commonly discussed in the TQM literature, customer satisfaction is the ultimate goal of TQM. Examples of organic type of control systems found in control systems literature are strategic interactive controls (Simons, 1995); control systems that provide broad scope information, flexible aggregations and integrative information and information provided in a timely way (Chenhall, 1986); and strategic control systems (Ittner & Larcker, 1997). Due to its strategic characteristics, it is clear that the implementation of SCS 'fits' with the ultimate goal of TQM which is customer satisfaction.

A case study undertaken by Andersen et al. (2004) confirmed that TQM could be implemented more successfully with the presence of SCS. From their findings, it can be concluded that although the reasons for failure of TQM are complex, the result that suggesting this failure relates to poor linkage between TQM and SCS was convincing. However, their study was narrowly focused on only one element of SCS which was performance measurement. Therefore, the researchers suggested that more future study is needed to consider wider SCS framework. Concurrently, as a case study, it naturally imposes a limitation of the generalizability of their findings. Therefore, a further study involving larger sample is needed to help validate and extend the generalizability of the findings.

5. Model of the study

The premise of contingency theory is that the strategy and control systems must somehow 'fit' together if the strategy is to be successfully executed. This premise has been widely investigated in the strategy literature as well as management accounting literature with the various operationalization of the concept of fit. Moving from the premise of contingency theory, this study postulates that the unsuccessful of TQM could be related to the misfit between TQM and control systems.

There are various ways of investigating the concept of 'fit' under contingency theory. However, this study has developed the research framework based on the concept of 'fit as mediation' as put forward by Venkatraman (1989). The idea behind the concept of 'fit as mediation' is the link between strategy and outcome mediated by the control systems. Figure 1 indicates the relationship between TQM and performance is mediated by strategic control systems.

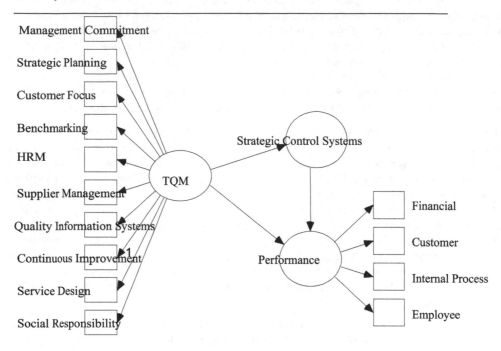

Fig. 1. A Structural Model Incorporating the Relationship between TQM, Strategic Control Systems and Performance.

6. Hypothesis development

After completing an extensive review of TQM literature, Sila and Ebrahimpour (2002) concluded that previous empirical evidence on the relationship between TQM and organizational performance had revealed inconsistencies in findings. In response to these conclusive results, Ehigie and McAndrew (2005) suggested that future researchers need to investigate variables that could influence the success of TQM. Among key variables that have been widely discussed in the literature and able to shed a light in explaining the relationship between organizational strategy and strategy related performance is control systems. In management accounting literature, the discussion about the control systems that is required for TQM to be successful has received much attention (Andersen *et al.*, 2004; Daniel & Reitsperger, 1991; Ittner & Larcker, 1997; Selto *et al.*, 1995).

The relationship between TQM and organizational performance is not restricted to the direct relationship; perhaps it could be indirect through the presence of SCS. As reported by

Horovitz and Thietart (1982), the use of the suitable control systems is among the prerequisite for a strategy to be successful. The implementation of TQM, for example, cannot be separated from organizational structure like control systems (Othman, 2000) since the organizational control systems have an important role in supporting TQM implementation (Moura E Sa & Kanji, 2003). As such, the function of organizational control systems is to monitor the development of TQM towards the predetermined goals.

Additionally, Chenhall (2003) claimed that the appropriate organizational control systems must exist to support TQM. In line with Chenhall (2003), Daniel and Reitsperger (1991) also found that the organizational control systems must be consistent with TQM to gain good organizational performance. Drawing from the findings of Daniel & Reitsperger (1991) as well as the premise of contingency theory, this study postulates that TQM can be implemented more successfully through the presence of TQM focused control systems. This structural relationship is also in line with the concept of 'fit as mediation' proposed by Venkatraman (1989). In line with the above discussion, this research hypothesizes:

H: The structural model of TQM and organizational performance is mediated by strategic control systems.

7. Research methodology

7.1 Unit of analysis

Unit of analysis of the study is the department of City Council (CC) and Municipal Council (MC) in Malaysia West.

7.2 Respondents

The respondents of the study were Heads of Department (HOD). They are most familiar with their departments' practices and performance results.

7.3 Population and sampling frame

Table 2 tabulates the number of existing departments by each city council and municipal council in Peninsular of Malaysia. The total number of existing departments is 341. However, 36 departments **(figure in bold)** were involved at the pilot study stage. Therefore, the balances of 305 departments were considered as the sampling frame for the main field work. The sampling frame of this study was developed by using two reliable resources, namely official websites of related local governments, and direct contact with officers from related local governments by using email or telephone. The latter approach was used due to the technical problem with official websites of related local governments during the data collection process. For example, Sungai Petani MC did not have an official website, thus a call was made to their Public Relation Officer. Kulim MC did not have such information eventhough they have their official website, thus a call was made to their Head of Administrative Department. Langkawi MC did not develop an official website, thus a call was made to their HOD of Management. Information about Teluk Intan MC was e-mailed by their officer.

Local Government	Existing Department
City Council	
Kuala Lumpur	22
Johore Bharu	7
Alor Star	8
Malacca	13
Ipoh	9
Shah Alam	12
Petaling Jaya	14
Municipal Council	
Batu Pahat	6
Johore Bharu Tengah	9
Kluang	6
Muar	7
Sungai Petani	10
Kulim	10
Langkawi	8
Kota Bharu	**8**
Alor Gajah	11
Seremban	**11**
Nilai	6
Port Dickson	9
Kuantan	11
Temerloh	13
Manjung	10
Taiping	8
Kuala Kangsar	7
Teluk Intan	8
Kangar	8
Pulau Pinang	10
Seberang Prai	10
Ampang Jaya	10
Kajang	**10**
Klang	10
Selayang	11
Subang Jaya	9
Sepang	7
K. Terengganu	**7**
Kemaman	6
Total population	341

Source: Developed by researcher based on information of every local government

Table 2. Existing Department by Local Government in Peninsular of Malaysia

7.4 Sampling procedure

This study used stratified cluster sampling. For groups with intragroup heterogeneity and intergroup homogeneity, cluster sampling is most appropriate (Sekaran, 2003). Generally, the characteristics of local governments in Malaysia are as follows:

- Governed under the same act, Local Government Act 1976
- Monitored by the same ministry, Ministry of Housing and Local Government
- Have similar functions, roles, objectives and types of activities as prescribed in Town and Urban Planning Act 1976 and Road, Drainage and Building Act 1974.
- Minimal difference in the organizational structure of each local government
- Many departments with different kinds of objectives, functions and activities within each local government. For instance, the Town Service Department and the Engineering Department.

Given the above characteristics, cluster sampling was applied for selecting samples. According to Davis (2000), cluster sampling is one of the methods more widely used in large scale studies. However, cluster sampling is exposed to larger errors than other probability sampling. This larger error occurs because the selection of each sampling unit within the cluster is dependent on the selection of the cluster, although the cluster is randomly selected. In order to reduce the loss of precision from cluster sampling, selection of cluster was stratified according to the status of the local government. This stratified cluster sampling (Davis, 2000) selects cluster at random from pre-specified strata.

8. Findings

8.1 Descriptive analysis of the constructs

According to the mean score, the implementation of each TQM factor and SCS are expressed as a high or low degree of TQM or SCS. In addition, the level of organizational performance is expressed as high or low performance.

Constructs	Mean	Standard Deviation	Min	Max
Management commitment	4.19	0.67	2.00	5.00
Strategic planning	3.85	0.62	2.00	5.00
Customer focus	4.00	0.60	1.00	5.00
Benchmarking	3.72	0.67	1.00	5.00
Human resource management	3.59	0.74	1.00	5.00
Supplier relationship	3.44	0.51	1.00	5.00
Continuous improvement	3.87	0.70	1.00	5.00
Quality information system	3.22	0.69	1.00	5.00
Service design	3.55	0.63	1.00	5.00
Social responsibility	4.03	0.68	1.00	5.00
Strategic Control Systems	3.66	0.61	2.00	5.00
Financial performance	3.67	0.74	1.00	5.00
Customer performance	3.62	0.61	1.00	5.00
Employee performance	3.67	0.66	1.00	5.00
Internal Process performance	3.63	0.69	1.00	5.00

Table 3. Descriptive Statistics of the Constructs (n=205)

As tabulated in Table 3, the mean value of management commitment is the highest among the TQM factors. This indicated that the commitment of management of local government understudied towards TQM was in a good situation. The minimum and maximum values of management commitment are 2.00 and 5.00 respectively, with the standard deviation of 0.67. However, the mean value of quality information systems is the lowest among the TQM factors with the value of 3.22. This mean value provided evidence that, more work needs to be done to improve the level of usage of quality information systems of Malaysia local governments. The possible explanation for this situation is that only a small number of local governments were advanced in their quality information systems.

8.2 Structural Equation Modeling test

SEM using AMOS was employed for examining the structural model of the study as depicted in Figure 2. SEM is an appropriate statistical technique for testing a model that is hypothesized to have relationships among latent variables that are measured by multiple-scale items, where at least one construct is both a dependent and independent variable (Hair *et al.*, 1998). As depicted in Figure 2, the TQM and OP are represented by oval which denote latent variable. TQM is represented as single latent variable composed of ten observed variables. OP is represented as a single latent variable composed of four observed variables. SCS is only measured by a single variable. In Figure 2, SEM attempts to account for random measurement error as represented by the small circles with the letter 'e'. The covariance matrix among the variables constituted the input for the SEM analysis. Although in practice both covariance and correlation matrices can be used as the input for SEM analysis, but the usage of covariance matrix is more recommended (Kelloway, 1998). By using AMOS package, the default matrix of this statistical package is covariance matrix.

The analysis of SEM using maximum likelihood estimation as performed in this study requires normal distribution of data. To satisfy this assumption, the test of normality, namely skewness and kurtosis were performed. As reported in Table 4, the value of skewness was between -0.427 and -1.293; and the value of kurtosis was between 0.205 and 2.768 respectively. Based on the results of skewness and kurtosis test, it indicated that the data of the study is within the acceptable level of normality assumption. According to Kline (1998), if the skewness is lower than 3.00 and kurtosis is lower than 10.00, the data has not violated the normality assumption.

The full model (input) is illustrated in Figure 2. The path coefficient between TQM and each of the ten indicators and the respective error variances were estimated, except that between TQM and management commitment that was fixed to 1. The path coefficient between OP and each of the four indicators and the respective error variances were also estimated, except that between OP and financial performance that was fixed to 1. On the other hand, none of the paths of the SCS variable was estimated since the SCS was explained by a single observed variable. Based on the suggestion by Kelloway (1998) in dealing with the condition of one indicator for a latent variable, the path coefficient between SCS and its error variance was set at a fixed value using the value of one.

Prior to testing the hypotheses, the model overall fit must be established (Bollen, 1989). The result of the full structural model with standardized parameter is presented in Figure 3. In order to evaluate the full structural model fit, a series of indices provided by AMOS were examined. Model fit determines the degree to which structural equation model fits the sample data. Model fit indices that are commonly used are chi-square (χ^2), goodness of fit index (GFI), adjusted goodness of fit index (AGFI), and root mean square residual (RMS). These indices are based on differences between the covariance matrix of observed model and the implied model (Hair et al., 1998). Among these indices, the chi square is the most popular index (Bollen, 1989). In evaluating chi-square value, researchers are interested in obtaining insignificant chi-square value. Generally, smaller chi-square value indicates a better model fit to the data. However, the index of chi-square has been criticized due to this index being very sensitive to the sample size (Bollen, 1989).To derive a conclusion, SEM literature is always suggesting researchers evaluate the model fit based on a range or series of fit indices. Table 5 depicts the list of goodness of fit measures and the levels of acceptable fit adapted from Hair et al. (1998).

Variables	Skewness	Kurtosis
Management commitment	-0.637	0.870
Strategic planning	-0.872	0.778
Customer focus	-1.037	2.274
Benchmarking	-0.777	0.740
Human resource management	-0.836	0.767
Supplier relationship	-0.683	2.062
Continuous improvement	-0.809	1.330
Quality information systems	-0.427	0.205
Service design	-0.604	1.262
Social responsibility	-1.293	2.768
Strategic control systems	-0.586	0.309
Financial performance	-0.694	0.628
Customer performance	-0.733	0.713
Employee performance	-0.880	1.274
Internal Process performance	-0.809	1.298

Table 4. Skewness and Kurtosis of Constructs

Goodness of Fit Measures	Levels of acceptable fit
Chi-square	$P \geq 0.05$
Chi-square/degrees of freedom	≤ 3.00
Goodness-of-fit Index (GFI)	≥ 0.90
Adjusted Goodness-of-fit Index (AGFI)	≥ 0.90
Normed Fit Index (NFI)	≥ 0.90
Non Normed Fit Index (NNFI) or Tucker Lewis Index (TLI)	≥ 0.90
Comparative Fit Index (CFI)	≥ 0.90
Root Mean Square Residual (RMSR)	≤ 0.08

Adapted from Hair et al. (1998)

Table 5. Levels of Acceptable Fit of Goodness of Fit Measures.

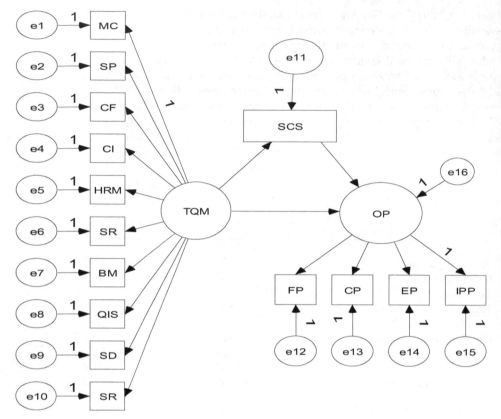

Fig. 2. Full Model (input)

However, the indices of the full model investigated in this study as given in Figure 3, did not achieve the suggested level. Thus, modifications on the model were done based on modification index. Modification index provides information for model modification. This step was redo several times until the model investigated fit the data.

Table 6 reports the indices of the modified model. The indices of the modified model surpassed or are marginally lower than the benchmark value suggesting the model did fit to the data. Given that hypothesis testing was evaluated. In order to test the hypothesis of this study, the significant of the structural paths were investigated by referring to the value of critical ratio (CR). A critical ratio is defined as the ratio between the standard estimation and its standard error (Arbuckle & Worthe, 1999). Normally, a structural path with CR value larger than 1.96 in magnitude is considered as significant.

The hypothesis of this study is the structural relationship between TQM, SCS and organizational performance fit to the data. Table 6 tabulates the index of modified model of this study had achieved the suggested level. Additionally, Table 7 presents the results of direct effect and indirect effect between the constructs under study. As can be seen, the total effect of TQM on OP is higher than the direct effect of TQM on OP. The total effect of

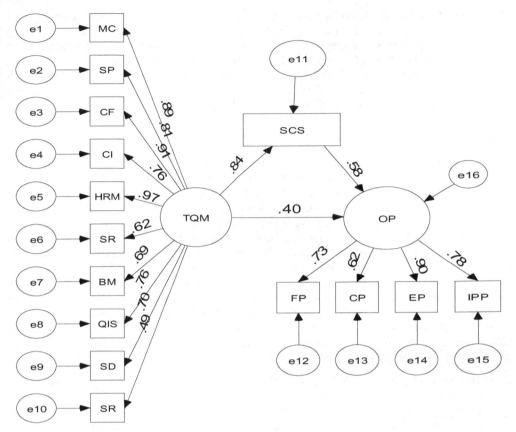

Fig. 3. Full Model with Standardized Parameter

Indices	Value	Threshold	Acceptability
Goodness of fit index (GFI)	0.848	≥ 0.90	Marginal
Relative fit index (RFI)	0.854	≥ 0.90	Marginal
Normed fit index (NFI)	0.902	≥ 0.90	Acceptable
Incremental fit index (IFI)	0.922	≥ 0.90	Acceptable
Tucker Lewis index (TLI)	0.833	≥ 0.90	Marginal
Comparative fit index (CFI)	0.921	≥ 0.90	Acceptable

Table 6. Indexes of Full Modified Model

Path	Direct effect	Indirect effect	Total
TQM→SCS	0.861		
TQM→OP	0.400	0.481	0.881
SCS→OP	0.559		

Table 7. Standardized Direct and Indirect Effect

TQM on OP can be calculated by adding the direct effect of TQM on OP and indirect effect of TQM on OP through the presence of SCS as follows. The indirect effect of TQM on organizational performance through the presence of SCS is calculated by multiplying the direct effect of TQM on SCS and the direct effect of SCS on OP ($0.861 \times 0.559 = 0.481$). Therefore, the total effect of TQM on organizational performance is 0.881. This result provides evidence that the impact of TQM on the level of organizational performance may improve through the presence of SCS. Therefore, H_1 is supported.

9. Conclusion

This study investigated the structural relationship between TQM, Strategic Control Systems (SCS) and Organizational Performance (OP). The investigation was motivated by the inconsistent findings concerning this relationship that appears in the literature, thus becoming another unresolved issue that needs to be scientifically revisited. Given the importance of control systems in executing organizational strategy, this study attempted to identify the mediating role of SCS on the relationship between TQM and OP. This study found that TQM on strategy through the presence of SCS had a stronger relationship with OP, as compared to direct relationship between TQM and OP. In other words, the finding indicated that the explanatory power of TQM toward OP was higher when mediated by SCS than that of TQM directly toward OP. Therefore, it could be concluded that the presence of SCS is essential to the success of TQM strategy. Perhaps, TQM strategy and SCS have a synergistic impact on OP. Therefore, these findings provide support for the earlier conclusion derived by Andersen *et al.* (2004). As concluded by these authors, an organizational strategy could be implemented more successfully with the presence of strategy related control systems.

The implementation of TQM needs high inter-functional activities (Groocock, 1986; Feigenbaum, 1986), whereby entire functional activities like research and development, purchasing, production, human resource, accounting and marketing are involved in the attainment of TQM strategy. In order to coordinate these high inter-functional activities, an effective control systems must be in placed (Zakaria, 1999). Based on the finding of this study, it suggests that local governments seeking to achieve better performance by institutionalizing TQM are subject to the effectiveness of TQM focused control systems. By having TQM focused control systems, TQM philosophy would be easier to be internalized by all employees as an internal culture of their organization.

10. References

Abdul Karim, M. R. (1999). *Reengineering the public service: Leadership and change in an electronic age*. Subang Jaya: Pelanduk Publications.

Ahire, S. L. & Golhar, D. Y. (1996). Quality management in large vs small firms. *Journal of Small Business Management, 34* (2), 1-13.

Andersen, H. V., Lawrie, G. & Savic, N. (2004). Effective quality management through third generation balanced scorecard. *International Journal of Productivity and Performance Management, 53* (7), 634-645.

Arawati, A. (2005). The structural linkages between TQM, product quality performance, and business performance: Preliminary empirical study in electronics companies. *Singapore Management Review, 27* (1), 87-105.

Arbuckle, J. L. & Wothke, W. (1999). *Amos 4.0 user's guide*, SmallWaters Corporation: USA.

Banker, R. D., Potter, G., & Schroeder, R. G. (1993). Reporting manufacturing performance measures to workers: An empirical study. *Journal of Management Accounting Research, 5*, 33-55.

Benavent, F. B., Ros, S. C. & Moreno-Luzon, M. (2005). A model of quality management self-assessment: An exploratory research. *International Journal of Quality & Reliability Management, 22* (5), 432-451.

Bollen, K. A. (1989). *Structural equations with latent variable*. New York: Wiley

Butler, A., Letza, S. A. & Neale, B. (1997). Linking the balanced scorecard to strategy. *Long Range Planning, 30* (2), 242-253.

Chenhall, R. H. (1986). Authoritarianism and participative budgeting: A dyadic analysis. *The Accounting Review, 2*, 263-272.

Chenhall, R. H. (2003). Management control systems design within its organizational context: Findings from contingency-based research and directions for the future. *Accounting, Organizations and Society, 28*, 127-168.

Chenhall, R. H. (2005). Integrative strategic performance measurement systems, strategic alignment of manufacturing, learning and strategic outcomes: An exploratory study. *Accounting, Organizations and Society, 30* (5), 395-422.

Cruickshank, M. (2003). Total Quality Management in the higher education sector: A literature review from an international and Australian perspective. *TQM & Business Excellence, 14* (10), 1159-1167.

Daniel, S. J. & Reitsperger, W. D. (1991). Linking quality strategy with management control systems: Empirical evidence from Japanese industry. *Accounting, Organizations and Society, 16* (7), 601-618.

Davis, D. (2000). *Business research for decision making*. 5th ed. Duxbury: USA.

Dent, J. F. (1990). Strategy, organizations and control: Some possibilities for accounting research. *Accounting, Organizations and Society, 15* (1/2), 3-25.

Ehigie, B. O. & McAndrew, E. B. (2005). Innovation, diffusion, and adoption of total quality management (TQM). *Management Decision, 43* (6), 925-940.

Emmanuel C., Otley, D. & Merchant, K. (1990). *Accounting for management control*. 2nd ed. London: Chapman and Hall.

Feigenbaum, A.V. (1986). *Total quality control*. 3rd. New York: McGraw-Hill

Goold, M. & Quinn, J. J. (1993). *Strategic control: Milestones for long-term performance*. London: Pitman Publishing.

Government of Malaysia. (1991). *Improvement and development in the public service*. MAMPU. Kuala Lumpur.

Government of Malaysia. (1992). Development Administration Circulars No. 3 of 1992: Manual on Micro Accounting Systems. Kuala Lumpur: Prime Minister Department.

Govindarajan, V. & Gupta, A. K. (1985). Linking control systems to business unit strategy: Impact on performance. *Accounting, Organizations and Society, 10* (1), 51-66.

Groocock, J. M. (1986). *The chain of quality: Market dominance through superior product quality*. Wiley & Sons: USA.

Hair, J. F., Anderson, R. E., Tatham, R. L. & Black, W. C. (1998). *Multivariate Data Analysis*. 5th Ed. Prentice Hall : USA.

Hayes, R. H. & Abernethy, W. J. (1980). Managing our way to economic decline. *Harvard Business Review, July-August*, 67-77.

Ho, S. J. K. & Chan, Y. C. L. (2002). Performance measurement and the implementation of Balanced Scorecards in Municipal Governments. *Journal of Government Financial Management*, 8-19.

Hoogervorst, J. A. P., Koopman, P. L. & Van der Flier, H. (2005). Total quality management: The need for an employee-centred coherent approach. *The TQM Magazine, 17* (1), 92-103.

Hoque, Z. (2004). A contingency model of the association between strategy, environmental uncertainty and performance measurement: Impact on organizational performance. *International Business Review, 13*, 485-502.

Horovitz, J. H. & Thietart, R. A. (1982). Strategy, management design and firm performance. *Strategic Management Journal, 3* (1), 67-76.

Hunt, V. D. (1995). *Quality management for government: A guide to federal, state and local implementation.* Milwaukee, Wisconsin: ASQC Quality Press,

Ibrahim, F. W. & Abd. Karim, M. Z. (2004). Efficiency of local governments in Malaysia and its correlates. *International Journal of Management Studies, 11* (1), 57-70.

Ittner, C. D. & Larcker, D. F. (1997). Quality strategy, strategic control systems and organizational performance. *Accounting, Organizations and Society, 22* (3/4), 295-314.

Juran, J. M. (1988). *Juran on planning on quality.* New York: Free Press.

Juran, J. M. (1995). A history of managing for quality: The evolution, trends, and future directions of managing for quality. ASQC Quality Press: Wisconsin USA.

Kanji, G. K. (2002). *Measuring business excellence.* Routledge Advances in Management and Business Studies. London, Routledge.

Kaplan, R. S. (1983). Measuring manufacturing performance: A new challenge for managerial accounting research. *The Accounting Review, Oct,* 686-705.

Kaplan, R. S. & Norton, D. P. (2000). *The strategy-focused organization.* USA: Harvard Business School Press.

Kelloway, E. K. (1998). *Using LISREL for structural equation modeling: A researcher's guide.* SAGE Publications, Inc: California, USA.

Khandwalla, P. (1972). The effects of different types of competition on the use of management control. *Journal of Accounting Research, Autumn,* 275-285.

Kline, R. B. (1998). Principles and Practices of Structural Equation Modelling. NY : Guilford Press (cited in) Widener, S. K. (2006). An empirical analysis of the level of control framework. Paper presented at the 2006 Management Accounting Section Meeting Conference (www.aaahq.org/mas/index.cfm).

Koehler, J. W. & Pankowski, J. M. (1996). *Quality government: Designing, developing and implementing TQM.* Florida: St Lucie Press.

Likert, R. (1967). The method of constructing on attitude scale (in) Martin Fishbein (ed) *Readings in attitude theory and measurement.* John Willey & Sons. Inc. 90-95.

Lin, Z. J. & Johnson, S. (2004). An exploratory study on accounting for quality management in China. *Journal of Business Research, 57* (6), 620-632.

Madu, C. N., Kuei, C. & Lin, C. (1995). A comparative analysis of quality practice in manufacturing firms in the U.S. and Taiwan. *Decision Sciences, 26* (5), 621-635.

Merchant, K. A. (1982). The control functions of management. *Sloan Management Review (Summer),* 43-55.

Merchant, K. A. (1985). Organizational controls and discretionary program decision making: A field study. *Accounting, Organizations and Society, 10* (1), 67-85.

Mohd Nizam, A. R. & Tannock, J. D. T. (2005). TQM best practices: Experiences of Malaysian SME. *Total Quality Management, 16* (4), 491-503.

Moura E Sa, P. & Kanji, G. K. (2003). Finding the path to organizational excellence in Portugese local government: A performance measurement approach. *Total Quality Management, 14* (4), 491-505.

Muhamad, M., Kamis, M. & Jantan, Y. (2003). Success factor in the implementation of TQM in public service agencies. *Analisis, 10* (1), 125-138.

Muralidharan, R. (2004). A framework for designing strategy content control. *International Journal of Productivity and Performance Management, 53* (7), 590-601.

Nor Hazilah, A. M. (2004). *Quality management in the public sector: An empirical survey of the Ministry of Health Hospitals in Peninsular Malaysia.* Unpublished PhD dissertation: Universiti Malaya.

Nunnaly, J. C. (1978). *Psychometric theory.* New York: McGraw Hill.

Nunnaly, J. C. & Bernstein, I. H. (1994). *Psychometric theory.* New York: McGraw Hill.

Othman, R. (2000). Contingencies of quality management approach: Evidence from the manufacturing sector. *Utara Management Review, 1* (2), 57-68.

Otley, D. T. (1999). Performance management: A framework for management control systems research. *Management Accounting Research, 10* (4), 363-382.

Otley, D. T., Broadbent, J. & Berry, A. (1995). Research in management control: An overview of its development. *Proceedings Vol 2. International Management Accounting Conference.* 19-21 January 1995. Universiti Kebangsaan Malaysia.

Powell, T. C. (1995). Total quality management as competitive advantage: A review and empirical study. *Strategic Management Journal, 16* (1), 15-37.

Prajogo, D. I. & Sohal, A. S. (2006). The relationship between organizational strategy, total quality management (TQM), and organizational performance-The mediating role of TQM. *European Journal of Operational Research, 168,* 35-50.

Prybutok, V. R. & Ramasesh, R. (2005). An action research based instrument for monitoring continuous quality improvement. *European Journal of Operational Research, 166* (2), 293-309.

Rangone, A. (1997). Linking organizational effectiveness, key success factors and performance measures: An analytical framework. *Management Accounting Research, 8* (2), 207-219.

Samson, D. & Terziovski, M. (1999). The relationship between total quality management practices and operational performance. *Journal of Operations Management, 17* (4), 393-409.

Sanchez-Rodriguez, C. & Martinez-Lorente, A. R. (2004). Quality management practices in the purchasing function: An empirical study. *International Journal of Operations & Production Management, 24* (7), 666-687.

Sekaran, U. (2003). *Research methods for business: A skill building approach.* 4th ed. New York: John Wiley and Sons.

Selto, F. H., Renner, C. J. & Young, S. M. (1995). Assessing the organizational fit of a just-in-time manufacturing system: Testing selection, interaction and systems models of contingency theory. *Accounting, Organizations and Society 20 (7/8),* 665-684.

Sila, I. & Ebrahimpour, M. (2002). An investigation of the total quality management survey based research published between 1989 and 2000. *International Journal of Quality & Reliability Management, 19* (7), 902-970.

Simons, R. (1987). Accounting control systems and business strategy. An empirical analysis. *Accounting, Organizations and Society, 12*, 357-374.

Simons, R. (1995). *Levers of control,* Boston: Harvard University Press.

Sinclair, D & Zairi, M. (2001). An empirical study of key elements of total quality-based performance measurement systems: A case study approach in the service industry sector. *Total Quality Management, 12* (4), 535-550.

Sohal, A. S. & Terziovski, M. (2000). TQM in Australian manufacturing: Factors critical to success. *International Journal of Quality & Reliability Management, 17* (2), 158-167.

Spencer, B. A. (1994). Models of organization and total quality management: A comparison and critical evaluation. *Academy of Management Review, 19* (3), 446-471.

Sunday Times. (2005). Written by Rostam, p.18

Terziovski, M. & Samson, D. (1999). The link between total quality management practice and organizational performance. *International Journal of Quality and Reliability Management, 16* (3), 226-237.

The Sun. (2004). News on page 1, 21 April 2004.

US General Accounting Office. (1992). *Quality management - Survey of federal organizations.* GAO/GGD-93-9BR, October.

Venkatraman, N. (1989). The concept of fit in strategy research: Toward verbal and statistical correspondence. *Academy of Management Review, 14* (3), 423-444.

Witcher, B. (1994). The adoption of total quality management in Scotland. *The TQM Magazine, 6* (2), 48-53.

Yasin, M. M., Alavi, J., Kunt, M. & Zimmerer, T. W. (2004). TQM practices in service organizations: An exploratory study into the implementation, outcomes and effectiveness. *Managing Service Quality, 14* (5), 377-389.

Zakaria, A. (1999). Budgetary Slack in Total Quality Management Environment. Unpublished PhD dissertation: Universiti Utara Malaysia.

11

Implementation of Quality Systems in Sociotechnical Systems

Mercedes Grijalvo

Polytechnic University of Madrid, School of Industrial Engineering,
Department of Industrial Engineering, Business Administration and Statistics
Spain

1. Introduction

In the increasingly dynamic environment which has been configured over the twentieth century based on the increasing demands of the environment, the need to innovate has been a constant. But if productivity was initially the key element to achieve, in the last quarter century, have appeared new elements where quality should be highlighted.

There is no doubt that these competitive pressures arising from the internationalization of markets that occurs at an increasing rate over the last decades of the twentieth century, have forced many organizations to implement quality approaches while the mismatches and errors arising from the application of traditional approaches to work organization, still not been sufficiently analyzed and diffused. A management approach based on hierarchy and control derived from an approach predominantly Taylors that is not always disputed by organizations in implementing quality models, despite its focus to prevention and organizational learning.

The reasons must be sought in the uncertainties and weaknesses of the models:

- *ISO 9000 standards* extend quality concept to areas related to the production process, but do not fall to discuss the paradigm of work organization to apply for the Organization (Lahera , 2004)
- The *EFQM model* adopts a systemic approach but it raises an additive scheme of interactions between its elements (Eskildsen et al, 2000; Conti, 2002)

As well as in the diffusion process, which has been linked to:

- ISO 9000 standards, to the demands of customers to implement and certify their quality systems, which calls into question their voluntary nature and the adoption of the approach to improving business continuity (Jones et al, 1997; Martinez, 2004).
- The *EFQM model*, to the adoption of an award requirements as criteria for implementing quality management, although their approach to evaluation (Zink and Schmidt, 1998)

This has resulted in quite a lot of deployments of these quality systems where become evident the contradictions and inconsistencies arising from the uncertainty of a paradigm of a clear underlying organization of work which leads us to propose a reflection about how it has been developed quality in different areas (in the West in general and Spain in particular).

2. Objectives and methodology

The aim of this study is to establish how quality systems models are influenced by work organization which can act as a limiting factor or facilitator and develop an analytical framework for the implementation of quality systems in organizations, which collects in an integrated manner the contributions of both disciplines, work organization and quality, in regard to socio-technical nature of the organization and the particular importance of quality as a learning tool.

To achieve these goals it has been developed a large study of the origins and the evolution that have been experienced as the work organization models, from the Taylors model to the sociotechnical systems model, as the quality management models, from the quality assurance standards and total quality models applied in the West and USA to the CWQC and lean manufacturing in Japan.

This bibliographic review examine in depth , in both disciplines, as the theoretical approaches and methodologies developed for its implementation, as the results achieved whit its application.

From this information on, a research work has been done in three large Spanish industrial companies to study in depth the relations and interactions between work organization models and quality tools and models through the evolution of the implantation process of quality management.

By means of all of this information, it studies and contrasts, comes up the connections between these matters and the identification criterions for the development and the implementation of quality systems in sociotechnical systems.

3. Development of the theoretical framework

3.1 Evolution of the industrial work organization models

The enormous success reached by USA in its industrialization process in the early twentieth century following the Taylorist model based on the division of labor and specialization of workers, made it became gradually extended in the developed world and become into the true paradigm of business management and organizations during the twentieth century.

Although the model had mismatches, and the proof of it are the results obtained by Elton Mayo at the Hawthorne plant, not until the early '50s when a series of British Tavistock Institute researchers call into question the Taylor paradigm as the only possible one, the one best way to achieve high productivity.

The Institute's researchers pointed out that the process of mechanization had destroyed the traditional structure of small groups of miners who carried out all of the tasks related to coal mining, replacing it with specialists who worked independently in highly specialized jobs.

The solution suggested by the researchers was a new form of work organization which they called the Design of socio-technical systems, inspired by systems theory, biology, logic, and cybernetics, from which they adopted concepts and theories which were later modified on the basis of practical experience:

- A holistic system, which makes it possible to look at the overall situation, adopting an integrated view of a production system made up of a technical system and a social subsystem in continuous interaction.
- An open system, where it is necessary pay attention to the relations between its elements and between the system and their environment.
- The self-regulation, which is the basis of work groups.

Conceptually, the new model means a change in the design of work organization. Traditionally, in the Taylorist model, engineers decide what form of organization is best, based on the demands of the technical system, and without taking into account the relationship between the technology and people. The theory of socio-technical systems, opposite to the Taylorist model, proposes the necessity of establishing a joint design of the technical system and the social system.

To do this, the socio-technical model proposes the analysis of both the social and the technical systems, on an equal basis, and the study of the relationship between them. This is because considerate that the proper performance of the system depends more on how its parts interact than on how they work independently and the optimum performance of a system cannot be reduced to the sum of the optimum performance of its parts (Trist, 1981).

An important paradigm shift in traditional Taylorist approach, which has its main exponent in the new role of workers within the company and which is reflected:

- Both in its design principles (Trist, 1981)
  - Endless process and minimum critical specification vs. the use of standards for control.
  - Multifunctionality vs. specialization in the work.
  - Sociotechnical criteria vs. separation in the planning and in the operation.
  - Location of the boundaries vs. functional organization.
  - Compatibility vs. Clear delineation of authority and responsibility
- As in applied practices for an implementation in organizations, carried out mainly through the working groups, which are integrated into what was called the operation unit, and are but different parts of process where the transformation of the product is made.

This recognition of the importance of the social system in the organization of work not only involves changes in the social system:

- The multifunctionality of the working groups allows the exchange of jobs, reducing the coordination needs between workers and bureaucracy and ultimately, increasing the autonomy of workers, especially in decision-making in those areas directly related to the activities they perform.

But also in the technical system:

- The new focus is on the product and not the task, on the outcome of the action, on the client, and not on the action itself.
- The objective is the control of the product by the workers, instead of the control of the workers.

An innovative approach, not nowadays but in 1950, which was created with the objective of improving life quality of workers, becomes equally valid to improve productivity and product quality and to facilitate organizational flexibility, achieving a double objective: satisfying the wishes of both workers and managers (Molleman and Broekhuis, 2001).

This view allows us to see once again the possibilities afforded by worker involvement and participation, from a different perspective. First of all, the workers share a common work language – the product – which makes them strive to achieve this one goal instead of simply doing the tasks that are assigned them without knowing what these are for. Secondly, they are in the best position to control the process while they are working: they know how the process works, what its key variances are, and how to keep these under control.

But the model despite the good results achieved in some industrial experiments, has had a very limited distribution, in contrast to other models and management techniques, such as in quality models (Jackson, 2003).

This fact and the strong dependence of the methodological approach of the model, the action research, not only have difficult its further development, especially in technical and methodological aspects, but which led to the continuation of the Taylor's model.

For though the prestige of the Taylor's paradigm has been declining to the point that few managers today would dare to confess that they are still applying it, the fact is that many of the basic elements of this approach are still applying it today, more than one century after his birth.

In this context, in the following lines are being examined the models and techniques of quality. In these management models can be observed how the absence in its definition of clear references to one or the other paradigm of work organization has made to be common to see in many of their deployments, clearly Taylorist elements with others that substantially differ.

At the same time and despite this lack of reference, in its evolution (these methods emerged from under taylorist environment) may find the continuity of critical socio-technical approach, turning to cast doubt on such principles as: the maximum division of work, extreme hierarchy, the excessive mechanization and automation, etc.

3.2 The role of quality in a global economy – Nature and importance of quality

3.2.1 The quality management – Technique, strategy or philosophy?

From the historical point of view, it can be said that it is in the late '80s when Western companies began to implement quality systems and techniques widely in response to the competitive challenge launched by Japan.

The absence of competitors made in the '50s that Western companies leave quality techniques and their interest focused on producing and selling goods for the world market

(Ivancevich et al, 1997). The companies, more concerned about issues such as improving productivity or the incorporation of advanced technologies (ERP's, etc.), forgot about the customer.

This overconfidence led to stagnation in the evolution of quality in the West, Japan spoke about ppm while the West was still talking about acceptable percentage of defects. While it is clear that the effort done in terms of quality in Western companies has recovered this focus on the customer, the quality has been developed from a highly technical form of the labor standards and models assessment and specialists in the field.

These systems, on the one hand, allow companies to both reduce costs by avoiding errors and time wastes and achieving improvements in business processes, as well as reaching a high degree of differentiation and reinforcing its brand image (Claver et al, 1999; Tari, 2001).

But this step forward in rationalization of the processes of quality assurance of the providers and also in spreading the culture of quality, poses risks, especially as companies face the pragmatic implantation, underestimating the cultural impact involved in its implementation (Camison et al, 2007).

And despite this implicit recognition that if you want to improve, is necessary to impact on the way of working on certain products, is also obvious that often influence on the design of the organization of work is avoided and assuring quality is limited to the control of the process by setting certain standards or guides of good practice that are slowly converging to international recognition models but are made outside the system itself which they are applied to (though often sectoral specificities are recognized, of organizational size, etc.) and to which are incorporated a certification scheme for its audit and evaluation.

Its rapid diffusion through the demands of customers requiring certified quality systems has prompted many companies to focus their efforts in meeting the requirements of the models, creating a dynamic of "ensuring the existence and functioning of the system", that has been left out not only many of the principles of quality management, but also the improvement of productivity.

The certification does not guarantee that the product quality is good, or that the quality improves with time, or that the products of the registered companies meet the needs of its customers, or the quality levels of products of companies registered is such or simply that the quality of these products is much better than those of non-registered (Ciampa, 1993; Rotger and Canela, 1996).

In this sense it is possible to say that this process of operationalization and instrumentation of quality, which seems that apart from continuing the approach of total quality management around the EFQM model and the Excellence Accreditation, has resulted in significant competitive advantages for companies, and in parallel has also fallen into another series of serious errors, because the implementation of quality required to put into practice a series of organizational mechanisms designed to facilitate and promote the participation of everyone in the organization in the control and improvement of quality (Prida and Grijalvo, 2008).

Behind this, is not but the difficulty to perceive, as different cases of companies analyzed, how old methods of motivation to prioritize the performance or use of standards for the control of those people still present in the culture of the organization, even years after they have been eliminated, acting as constraints to change (Prida and Grijalvo, 2007; Prida, 2005).

Many of the problems of implementing these models arise when procedures and standards, key elements to achieve the stability of the process and therefore also to apply improvement techniques, are used exclusively for the control of labor that has been excluded from any real participation in their design and implementation (Lahera, 2004).

Errors must be detected and managed there where they are produced and this is not possible without the participation of those who do the work. Empowerment is not a new mechanism for motivating workers to accept executing that which is derived from plans designed exclusively by management, but is a necessary tool for the management and learning of the organization starting from their base.

The Japanese culture was the pioneer in mobilizing human resources from improving quality, enhancing mechanisms for employee participation. The involvement of employees within the process becomes a common form of development work in Japanese companies. The goal is to delegate as much planning and control as possible, making workers responsible of the quality and the improvement (Imai, 1991).

It is not that workers participate in strategic decisions of the company but constantly studying the problems that appear within their immediate environment and solving them, so in this way, they constantly improve the results of their work and the environment in which they play it.

This brings as a result the redefinition of the roles of all personnel of the organization. It contributes significantly to the redesign of the organization, not so much on its organizational structure, which maintains the traditional hierarchical-functional, but the design of work organization, for initiatives to promote interactions between the different units and teamwork (Tillery and Rutledge, 1991; Imai, 1991; Beckford, 1998).

This is a very interesting aspect, given on one hand, the existing debate in the West about the need for changes in organizational structure to implement quality management and, in parallel, none of the models mostly used by companies: the rules ISO 9000 and EFQM Excellence Model, explicitly contemplate.

In this sense Belohlav (1993) and Conti (1993) assert that the problem in the implementation of quality management is not so much that the company is organized according to models of work organization based on hierarchy and standardization of activities, but in non-structural activities that the company carries out to ensure coordination between various departments and to break down any barriers to effective communication processes.

The evolution of the standards of assurance quality management in its review in the year 2000 has been marked by the introduction of these approaches through the principles of quality. While it should be noted that the solution is not so much in the fact that the models adopt these principles, but on the development of policies and guidelines within companies that create a culture of commitment, participation or improvement at a medium and long term.

Principles and practices generally are not independent, but are correlated and the success in the implementation of quality models and techniques depends largely on achieving this synergistic effect (Gown et al, 2007). Otherwise you may encounter problems and practices implanted, instead of exercising a facilitating or driving role in the implementation of the principles of quality that may act as a limiter or brake.

Although there are no explicit references to models of work organization in the techniques of quality, research has shown that the continuity of a critical thinking about the socio-technical approach in the Taylorist model in Japanese approaches, the CWQC and lean manufacturing, that return into question principles such as: the highest division of labor, extreme hierarchy, the excessive mechanization and automation, etc. At the same time it has drawn attention to the validity of Taylor approach in some of the existing quality models and whose most extreme case is probably the re-engineering of Hammer and Champy.

3.2.2 The systemic nature of principles and quality practices: Neo-Taylorism or continuous improvement

After this analysis it can be said that this new way of understanding human resources management has resulted to be the key strategy in the implementation of quality systems in Japanese companies, whose products in the 80's burst massively displacing many Western products of its own common market.

The KAIZEN integrates all the concepts and methodologies developed by American scientists and later by the Japanese Ishikawa, Taguchi and Shingo (Figure 1). All are tools and techniques to improve quality and do not have, as raised in the West, an end in themselves, but represent efforts of improving of all personnel in the company in different areas (Butch and Spangler, 1990; Ishikawa, 1988; Walton, 1986).

This is because of the approach to the system adopted by Japanese companies in their implementation; where so important are both the technological elements of the model and the humans', and the interactions between them (Imai, 1991; Mizuno, 1992).

Simple tools such as Kankan or suggestion systems have taken 10 years for its adoption in Japanese corporations. The implementation of these techniques requires time not only for training people in new skills, but also for undertaking them and putting them into practice (Galgano, 1993; Fortuna, 1991; Camison et al, 1991).

In addition, productivity and quality are inseparable linked (Coriat, 1993). The idea is to design systems that allow more dynamic response to an increasingly demanding and changing environment, not only in regard to aspects of quality but also to those related to service: responsiveness reliability, speed deliveries and the own flexibility to adopt to the changes.

In this sense, JIT is a technique that is based on the application of continuous improvement and waste elimination but whose implementation directly affects the production environment and also means to change the focus of cost management systems and the criteria for its measurement and control (Mcllhattan, 1991).

And is that not only now the cost management system should aim to identify the real usefulness of added value activities, but also that some traditional measures of cost accounting systems can encourage contrary actions to the desired ones: the extent of the use of machinery can induce the production of a good, creating inventories with an advance of the needs.

Fig. 1. The KAIZEN umbrella (Imai, 1991)

Therefore, Japanese companies do not separate the quality of the other demands on the system, but they integrate all these aspects to achieve a more robust system that is capable of responding effectively and efficiently to the demands of the environment in general and customers in particular.

Systemic approach is probably where the own approach goes far beyond the specific techniques, what has not always been understood in the West. This has facilitated the spread of many of them, but has meant that many companies understand them as simple methods to improve productivity that can be applied in a pragmatic way.(Jackson, 2001).

So is common to find organizations where they try to copy imitating these techniques, where these and objectives are confused or where the great philosophies can last for months before being replaced by new ones.

This utilitarianism and fragmentation, whose origin is in the ignorance of the functioning of systems with socio-technical nature, and that is the cause of many of the failures that occur in its implementation, can also be observed in the application of total quality models where again the West copies Japanese practices, in this case the Deming Prize, and where again is made the mistake of applying them in a pragmatic way.

For Nieto and Ros (2006) this award of excellence, the first that was created, although is based on non prescriptive guidance, presents a more technical nature with regard to the considerations to take into account when implementing the CWQC model:

- Both regarding to its principles, which are based on Deming's 14 principles
- as on the criteria for granting, which are based on statistical control in problem solving and continuous improvement.

By contrast, the criteria of EFQM model and Malcolm Baldridge are the criteria to examine, evaluate and rate those organizations wishing to present to the awards, and therefore, what they reflect is the opinion of several experts on the philosophy, methods and issues that, in their opinion, the organizations should take into account to address management excellence.

Cases such as Wallace & Co. firm, released even by the lay press (Ivey and Carey, 1991) that few months later after it was awarded with the Malcolm Baldrige American excellence prize, finished by disappearing because of bankruptcy, but can be considered as a anecdote that shows the need to relativize the word excellence in this type of model.

Conti (2007), one of the fathers of the European model of excellence, the EFQM, had already pointed in this direction some of its weaknesses in its systemic structure, pointing out its limitations as a benchmark for improvement.

But although the model recognizes that the company is a system and its various elements are interrelated, it does not explain how these interactions occur; that is to say, which is, in general, the common relationship between enablers and results, but taking a global approach: through innovation and learning is enhanced the work of the enablers to lead to an improvement in the results.

In addition, they establish a point model to evaluate each part of the system, without take in consideration the interactions among the various stakeholders which are operating in an organization, instead of the fact it is well known that in a system the sum of the parts cannot explain the behaviour of the whole.

This ambiguity of the systemic structure which the model presents, determines its usefulness to establish concrete plans of action in complex systems of socio-technical nature. The complex network of feedbacks of this type of systems against overly simple structure of cause-effect relationships established by the model, can lead the organization to make mistakes, thinking that the best way to achieve excellence is to focus on actions to add the maximum amount of points in the model or undertake less problematic actions to quickly demonstrate improvement.

In this sense, excellence models give indications according to the ultimate goal to be achieved by the organization to achieve the highest score on the excellence, but in them there are little evidences regarding to the best way to reach it.

Is its development, which is linked to the prize, what makes its application within the continuous improvement cycle to be centered almost exclusively on one of its phases, the evaluation, forgetting the rest, what certainly limits its usefulness as a learning tool.

On the other hand, the evolution of the ISO 9000 of quality assurance has been marked in its review on year 2000 by introducing approaches to improve quality through quality principles: focus on customer, leadership, involvement of the staff, approach based on the processes, system approach for management, continual improvement, factual approach for decision making and mutually beneficial relations with the supplier.

But implementing these principles goes beyond providing quality products or services and following predetermined procedures. It is a new way to organize and to work in the company: barriers between departments and functions should be break down, providing visibility and transparency, improving workflow and communication, etc.

The implementation of the standard requires, therefore, to consider three dimensions: technique, people and organization, since the mere documentation of practices that are taking place in the company, which may have even incorporated the requirements established in the rule, all it does is its formalization.

However, numerous studies show that all these aspects are not always addressed by companies in the process of implementation of standards. So, to the extent that the rule only states what the company should do and does not indicate anything about how to do it, the application of approaches to processes, continuous improvement and equal participation of workers, is strongly influenced by factors such as : the commitment of the company, its organization, its dynamic of change, etc. (Muthler and Lytle, 1991).

The reasons which may be behind this situation can be varied but in general they are all directly or indirectly related to the acquisition of the quality system certification, one of the objectives, if not the most important of those persecuted by the companies when they decide to implement a quality system according to international standards (Nanda, 2005; Celada, 2005).

Although the companies also hope to achieve with the implementation of standards-benefits related to the improvement of the quality, of both products and processes that lead to be more efficient, certification is the key element that provides value and that converts system of quality management into a competitive advantage (Duran, 2005).

Special mention, within the quality models based on standards, should be made of two sectors such as the automotive and the aeronautics where the pressures of big corporations, leaders in the sector, have led to the fact that the implementation and certification of the quality system is not anymore a differentiating factor but an essential requirement to operate (Grijalvo y Prida, 2005).

In this situation the need to facilitate both the implementation and certification of the system can lead to avoid, as far as possible, changes in the organization and processes to meet the requirements of the standard, because it allows them to both reduce efforts and reduce delays (Martínez, 2005).

Perhaps because companies are not sufficiently aware of the conflict of the principles of quality that are intended to implement with some of the models of work organization derived from Taylorism. For example:

- If it is accepted that there is one best way of doing the work, continuous improvement is meaningless.
- The only activities in which direct workers can participate in the Taylorist model are the ones about implementation that have been assigned by the technicians.
- The technicians' activities, in contrast, are reduced to planning, preparing and controlling, and are based on the exercise of authority and hierarchy, not leadership.
- The extreme division of labor looks for the decomposition of the process in its elements and thus, for the improvement of efficiency and effectiveness, are focused on each of them independently, forgetting that the organization is a system and the importance of the operation of the interactions between them and their environment.

In contrast, the models derived from the model of socio-technical systems design of work organization, fit much more to these new principles of quality. For example:

- The focus on the system so that the results depend both on the different elements and on the interaction between them and the environment, what provides a stronger basis on which to define the processes, their interactions and their improvement.
- The focus on the participation and worker self-control, which allow more efficient monitoring of the process' variability and establishes the necessary basis for the connection between processes.
- The flow of information, aimed to those who must take decisions, have the data necessary to do so.

Therefore it is necessary to prevent the implementation of the rules against this false sense of ease, particularly attractive to managers looking for quick results. The speed is not consistent with the slow pace of cultural change which takes a long time to involve people in the use and improvement of the system as part of their daily work.

The cases of companies that are discussed below show:

- The implementation of quality management requires the company to adopt a contingency approach that allows maintaining internal coherence between the different areas of management. The transfer of management techniques is not a linear process of diffusion, but a process of adaptation of those involved in the context where it is being implemented.
- The relationships and interactions between models of work organization and the techniques and models of quality through the development process of implementation of quality management, proposing different ways to address improving the competitiveness of the organization, but always based on the perception of the role of workers in the control and quality improvement.

4. Business case study

The study of these cases has allowed setting the evolution of the processes of implementation of quality management in enterprises, as well as the relationships and interactions that are established between the models and techniques of quality and organization work.

The choice of this type of qualitative methodology to carry out this stage of the research meets the objectives of our work, since the aim is partly to deepen the relationships and interactions between them, not to establish a statistical measure in the application of certain practices and on the other hand, to develop a model or framework of analysis with methodical recommendations for the implementations of the techniques and models of quality companies (Rialp, 1998; Eisenhardt, 1989; Glaser and Strauss, 1987; Ying, 1989).

The three selected companies belong to the industrial sector, more specifically to the automotive, the agricultural and the aerospace, have all a long way in the implementation of quality models, although it is possible to find significant differences between them:

- In the three companies manufacturing flawless products has always been an imperative for their business and in the three of them focus on quality assurance based on a mechanistic model, has been insufficient to achieve customer satisfaction, while the new approaches taken by each of them are different both from the standpoint of to the organization of work.

- In the three companies to move towards greater complexity either in the products made, in production equipment or in both, has not only but increased the importance of the participation of human resources, although this process has been tackled in a different way.

The study has shown that though the models in both areas, quality and organization of work, evolve in a structured way over time, the path taken by the companies, which otherwise reflect different models analyzed in this research:

- The first case, the automotive sector company, presents an alternative closer to the one developed by Japanese quality model. The commitment of the management of the company with quality is geared within the company to train workers in technical problem solving and teamwork, to subsequently be able to manage their own processes, after which they will be available to participate in defining the mission and vision of the company.
- The other two cases, companies in the agriculture and aerospace sector, pose an alternative closer to the developed by Western models of quality. The commitment of the management of the company with quality is geared within the company to define the mission, vision and long term objectives, which are then deployed by the organization, analyzing the critical processes for their achievement, for which, now, a technical staff is in continuous improvement, teamwork, etc.

They are different ways to address improving the competitiveness of the organization, but are based on the perception of the role of workers in the control and quality improvement. It is this conviction that they all lead to the development of organizational or social innovations, it lead to a clear empowerment and to a participation of workers in decisions affecting the organization of work.

4.1 Total quality as the development of organizational culture – The *"mini-companies"*

The company is one of the 19 factories of a multinational automotive sector and since its foundation it has focused its activities mainly in the manufacture of thin cables and wires for the reinforcement of radial tires. The first years of operation of the facilities are very complicated and the results are not within expectations. Inefficiencies in production and especially the problems with the quality of the products manufactured for its main client, created much discontent among both managers and employees.

In this context, managers "took the decision of concentrating all their efforts on quality and use the total quality management as the most appropriate way to redress the operation of the enterprise" (Martinez, 1993).

The managing director asks his engineers to visit other plants and study how they operate. There they will get familiarized with the techniques of problem solving and teamwork and as they return they will develop and implement the technique of the Color Lines.

This is the first of a long series of techniques and tools that have been implemented in the company to support the improvement of management, two of which are discussed below delving into the reasons that have made that the effort to improve has been effective.

The Color Lines has an approach similar to quality circles applied in Japan and as them, are aimed to the participation of people, with facts and data management and the systematic approach to improvement.

In general, each Color Line refers to the entire production process of a product or a small family of products and integrates all the people who have relationship with it. Each Color Line has established key parameters of which information is collected on daily activity and that are followed up in monthly or bimonthly meetings.

This methodology is applied initially to a small process, which is called "Orange Line", where many problems were detected, while the dramatic improvements in results in this line in a short time served to initiate a process improvement throughout the organization led by its managing director that decided to "do whatever was necessary to provide engineers and workers the resources, skills and tools to start the journey with no end in continuous improvement."

And that is translated into a series of training initiatives focused not only on providing to the plant personnel an awareness of quality concepts and practices, but above all, to develop especially an attitude of learning that can facilitate the development and deployment in the enterprise.

Therefore, the "Color Lines" are considered the basis for the development of values and culture of quality (Club Gestión de Calidad, 1995). Its application integrates many technical and social aspects to promote active employee participation and collaboration among workers and staff of the service areas in improving the productivity of processes. A comprehensive approach not only allowed changing the rules but also the organizational skills to make possible a system in which workers can better control the variability of work.

This gradual and systematical increase of the responsibilities of workers associated with the implementation of these and other quality techniques and tools, leads to the development of a new model of work organization, the "mini companies", a model for the autonomous management of small business units, a colored line or part of a color line, with common activities and objectives.

Each "mini company" is made up of customers, suppliers, bankers (Chiefs) and its members (employees), functioning as a real company with its own mission, its customer-supplier letter and its communication channels and it defines and deploys annual improvement plans based on their objectives (Suzaki, 2000).

The management of its daily activity is carried out through two cycles SDCA / PDCA: standardize, find problems, solve problems and implement new methods. A methodology aimed at improving outcomes in Safety-Quality-Delivery- Cost-Moral, the "SQDCM."

Despite this, the key for the development of self-management in "mini companies" has been the continued development of the skills of workers which has been promoted from the "Respect for people," one of the values on which it bases its culture and involves considering all workers, not as a simple resource or manpower but as human beings who seek self-fulfillment through their work. That is what they call solve "a problem feet" (the distance between the head of the heart), and it brings to contemplate the whole person: hands, brain and heart.

In this sense, is ceaseless to stress that this increase in responsibilities of employers was carried out gradually, what has enabled not only the formation of workers but the managers, especially supervisors, to explore new responsibilities without clinging to the current preventing the change.

Overall the experience of the company in the extension of the model within the enterprise shows that "Only those who have internalized the basic principles that underlie the systems have further success in the application in other cultures. It is not a matter of copying *the know how*. You do not get sustainable and lasting improvements without internalize *the know why*. "

4.2 Continuous improvement in business processes – Work teams

The company is one of the factories of an American enterprise and whose facilities were initially used in the production of agricultural machinery for the Spanish market, and later specialized in the manufacture of components for further assembly factories of the group in the world.

The evolution of quality in the factory reflects the perseverance in improving the quality of the products for its customers, which has been accompanied by an integration project of the human resources in the production processes, which has promoted major changes in the organization work (Dueñas, 2010; Macías, 2010).

At the end of the '50s, the initiatives around quality in the company were focused on quality control approach orientated both to standardization or development of technical specifications as well as to the use of control techniques and inspection of products and processes.

In addition, until the '80s the company applies traditional models of functional and hierarchical work organization based on the division of labor, and also on time and movements study, the basis of the pay system of the factory, the incentive system.

This pyramidal structure allows the company to grow by introducing unqualified people or with very simple levels of qualification, but the increased complexity of the factory and the loss of visibility of material flow through the different specialized departments makes necessary the redesign of the structure of production processes.

The factory is geared towards the product rather than to the manufacturing processes throughout the cellular manufacturing and, therefore there, appear the first initiatives for the versatility of the workers, who must now operate several machines.

The need to operate in the new market where the plant will be open to, that happens from producing a single final product to produce a variety of components for other plants, results in the mini factories, specialized production areas, each of which fully addresses the development of a separate component, and whose implementation involves the decentralization of some services such as production control, which passes directly to depend on these new units.

Since 1997, the company addresses the improvement of its quality system through the implementation of different models of quality: ISO 9001:1994 and the Corporate Quality Manual.

In parallel the company will implement the Teamwork, model based on methodologies of group work that is associated with an enrichment of both the vertical position by decentralizing decision-making, and the horizontal promoting versatility versus specialization and which seeks to respond to the new technological level of the plant with production lines becoming increasingly automated and characteristics which requires some characteristics and superior skills of the operators.

Despite this, the participation of workers is still very limited in spite of their participation in the Quarterly Cycle of Continuous Improvement, a methodology that consists of several phases based on the model of improvement of 6-sigma improvement, DAMAIC cycle, very similar to PDCA cycle

Meanwhile we must consider that creating in workers a positive attitude towards issues such as implication and assumption of responsibilities is a process that requires a relatively long time and that passes through a realistic assessment of structural inertia that programs on change must meet. In this sense, it is important to note the revised wage policies, despite the difficulty of changing the incentive system deeply rooted in the organizational culture, to the fixed portion of salary or base now it must be added a variable that encourages:

- the quality over quantity, substituting the premiums to productivity for premiums for good parts and
- teamwork and cooperation over the merits and individual skills, when replacing the right to collect the premium as the result of individual work for its collection according to the results of the work of the whole team.

The implementation of the process approach in the factory which started in 1998/99 with the implantation of Corporate Quality Manual, continues with the implementation of a guide developed by the Company to improve business processes, in which techniques are applied and also tools of six-sigma approach, the Green Teams.

The Green Teams is the first of several corporate assessment models whose goal is to standardize processes in different plants in the world that are focused on the implementation of six-sigma tools and techniques and, read in production processes, with which the Company looks not so much to introduce new requirements but to achieve better results.

The Models, in similar way to other models of self-evaluation, look both, for each procedure, identifying the strengths and areas for improvement of the plant, such as assigning a score. Noted that plants are not only not free to implement the model but also from the Company it is stated that they must obtain a minimum assessment to remain being suppliers.

4.3 A strategic approach to quality – Plant management areas

In this case it is presented and analyzed the case of a Spanish multinational (Grijalvo and Prida, 2006) from the aerospace sector which in recent years has undergone trough significant changes as a result of the following factors:

- The increasing globalization of the industry, which has significantly changed the characteristics of its market and has forced the company to cope with an increased international competition.

-   The change in ownership from public to private sector, which has been a major internal restructuring with the introduction of different foreign partners.

The whole process of deep changes in the company coincided with the increasingly determined incorporation of new approaches of quality in its management which over recent years, have not only been maintained but have renewed their impulse and no doubt that have certainly contributed to a competitive company today.

The recognition of the strategic importance of quality dates, in most sectors, of the last decades of the twentieth century, but in this case, the existence of departments of inspection and verification and the use of quality manuals have been integrated in the production activities, since much more years and manufacturing products with the guarantee of being free of defects, has traditionally been a business imperative.

It is noteworthy that in this enterprise customer requirements and the application of stringent quality standards, in some cases derived from the military sphere, joined to the high technological complexity of manufactured products, required the use of formal systems and quality manuals highly developed. In this situation the application of ISO 9000 initially assumed no more than minor adjustments in the systems and pre-existing quality manuals.

However, at the end of the 80's started to be felt in the company that this approach to "quality assurance" applied so far was insufficient both to achieve full customer satisfaction and to achieve economic efficiency in production operations, which facilitates the search for new ways to combine the new demands of the market with the usual quality.

The successful experience of a project to integrate human resources and material in production processes and management of factories in what is called Participatory Management coincides with attempts to move towards new models of total quality in the company and boost a new culture of continuous improvement in the organization that has promoted important changes in regard to its organizational structure and in relation to the management system itself.

Participative management is an organizational model that is based on the application of the following practices:

-   Organization by products and processes, to achieve simplification and optimization of processes.
-   Coordination / integration of the functions / activities at the lowest level, so that operational decisions are taken at that level.
-   Team work and multi-functionality focused on the process / product.
-   Dynamic adaptation to each situation through changes brought by the application of reengineering techniques and / or continuous improvement.

The aim is to increase people's participation in decision-making (within their level of responsibility), to achieve continuous improvement in the performance of the functions and activities assigned to each person and guide all efforts to the product to satisfy the customers.

For this, the company develops the Plant Management Areas. A Plant Management Area is a multi-functional team on a permanent basis, constituted within an organizational unit, where are, directly, effectively and in an integrated way, assumed and coordinated all activities associated with the process variables necessaries for the obtaining of a product.

For the implementation of Plant Management Areas, it is initially developed a parallel organization to the existing one, with teams around the production processes with hierarchical and nonhierarchical relationships among its members, to join them later when is changed the organizational structure of the company.

Plant Management Areas become a facilitator of the transition from functional hierarchical organization and the new organizational model adopted by the company which is the organization's processes; however, it should be noted that there are elements that can affect the development of the proposed change both from the social system and from the technical system:

- While the participation of people in the preliminary experiments was high, the new model involves changes in compensation systems which had been applied for years (more oriented towards functional productivity increase than customer satisfaction), which may give rise to situations of demotivation.
- The degree of repetitiveness and variety of manufactured products is not equal in all the production processes of the company and therefore, will not always be possible to propose the same solutions in all processes. On the other hand, it cannot be expected the same degree of control in high repeatability processes than in others that are ad.hoc processes established almost like a project.

In addition, the new organizational model marks a significant cultural change in the company, because although it was initially implanted in Manufacturing Directorate, also affects other units involved in the processes. This cultural change is also the main difficulty to which the company has to face with in its introduction:

- External orientation towards the client versus internal orientation towards the product.
- Results commitment rather than compliance.
- Process and equipment versus departments and heads.
- Participation and support against hierarchy and control.

The challenge now for the company is moving forward in organizational change implemented in production and in their new culture, extending it to the entire organization and in this line have been focused new improvement projects (Grijalvo and Prida, 2007):

- Both in the area of work organization, with the revision of the wage policies of the technicians and managers of the different divisions to align them with the group and the development of a new remuneration system which now includes a variable pay based on the fulfillment of objectives of the Company.
- As for quality, the company currently is doing in all of its plants, a project to implement lean methodologies in the production processes of all plants

5. Study results and conclusions

The study has proven that the success in implementing the principles of quality is contingent on the system of work organization in the company implemented and how it also strongly conditions beliefs and values shared within the organization, its culture (Ciampa, 1993, Merli, 1995; Camison et al, 2007).

Probably the competitive pressures that have come under Western companies since the 80's, have made much greater emphasis on customer focus and the structural changes that were needed to put it into practice that on policy human resource management related to work organization.

The substantial decrease in research activity in the organization of work in the design of sociotechnical systems in the 80's, whose proposals did not exceed the scope of experimentation, is but a reflection of the concerns of Western companies that have abandoned this focus of interest.

In contrast with the West, in Japanese firms, their concern to improve the quality was the trigger for a much larger systemic process that is reflected in the development of many techniques grouped under the name of Kaizen (Imai, 1991) and whose application results a change in organizational culture that involves the continued acceptance by all employees that their work is not only to meet the standards but also to improve them.

And most important, in general, its application results in a change in organizational culture that involves permanent acceptance by all employees that their work is not only to meet the standards but also improving them , with an approach that goes beyond the "zero defects" is not so much to get it right the first time, but doing it better.

Two different approaches related to the implementation of approaches to work organization that not only help to understand the contradictions in the implementations of quality systems in Western companies but, that companies do not obtain the expected results.

And is that traditional approaches of work organization based on hierarchy and control (Taylorism) or external motivation to work (School of Human Relations) are not always questioned by organizations when applying models quality, despite its focus to prevention and organizational learning. This makes that some companies:

-    Associate quality basically to compliance with the product features and dissociate it from productivity
-    See continuous improvement more as a technique than as part of their culture.
-    Raise the participation more as a necessary element for the process control than as a key element for continuous improvement.

In this sense, in Japanese companies implementing quality has always been closely linked to improving productivity and to the application of techniques of work organization, which affect both the system design, as well as the participation of workers in the process management and that has given rise to a philosophy that transcends the techniques used: focus on processes, internal customer, continuous improvement, etc..

Moreover, the empirical work has shown that the evolution of the quality system is not dependent on the application of certain approaches: quality assurance or total or specific models, EFQM and ISO 9000 standards, but whether they are linked to the restructuring processes to guide the client and the restructuring of work organization (Senge, 1995, Moreno-Luzon et al, 2001). It ultimately consists on enabling the participation of workers in the control and quality improvement.

It is therefore necessary that organizations address the implementation of quality from a socio-technical approach by applying models to improve the quality and organization of work to facilitate the participation of workers in management processes (Deming, 1989; Imai, 199). An integrated input from both disciplines, work organization and quality, in regard to the socio-technical nature of the organization and the particular importance of quality as a goal and as learning methodology, based on:

-   A strategic approach to quality through increased involvement of company personnel and the adoption of forms of work organization that give more power to workers in decision making or empowerment, that enables the management system to evolve over time regardless of the quality management model adopted.
-   An approach to the processes by redesigning the production system from two key elements: the structure of production and control structure that allows an approach to the complete structure of the system without going into the distinction between the social system and the technical system.
-   An approach to standardization and continuous improvement through the implementation of action research methodologies and the PDCA cycle, allowing an approach to more balanced technological innovation and linked to social development of the organization.

Is interesting to highlight the compatibility of this approach with the principles of quality and the ISO 9000 and EFQM models. Although, as has been noted throughout this paper, the use of references is a useful tool, its use and mainly its abuse involve risks, especially as companies underestimate the impact that involves the application or that this is done in a pragmatic way. They are not the principles or models that make an organization excellent today but are its structure and function the ones that allow adjusting to a changing environment.

6. References

Ackoff, R.L. (2002): El Paradigma de Ackoff. Limusa. Méjico.

Bailey, J. (1983): Job Design and Work Organization. Prentice Hall International, London.

Beckford, J. (1988): Quality: A critical introduction. Routledge. London.

Belohlav, J.A. (1993): Developing The Quality Organization. Quality Progress. October, pp. 119-122.

Buch, K.; Spangler, R. (1990): The effects of Quality Circles on Performance and Promotions. Humans Relations, vol. 43, n° 6, pp. 573-582.

Camison, C.; Cruz, S.; Gonzalez, T. (2007): Gestión de la calidad: Conceptos, enfoques, modelos y sistemas. Pearson Educación. Madrid.

Claver, E; Llopis, J.; Tarí, J.J. (1999): Calidad y Dirección de Empresas. Civitas Ediciones, S.L. Madrid.

Club Gestión de Calidad (1995): La formación para desarrollar actitudes favorecedoras de la calidad total: La experiencia de Ubisa. Capital Humano. VIII, Sep., pp. 22-24.

Ciampa, D. (1993): Calidad Total: Guía para su implantación. Addison-Wesley Iberoamericana. USA.

Cherns, A. (1976): Principles of Socio-Technical Design. Human Relations. Vol 29, n° 8, pp 783-792.

Conti, T. (2002): Human and social implications of excellence models: are they really accepted by the business community? Managing Service Quality, Vol. 12, n° 3, pp. 151-158.

Conti, T. (1993): Building Total Quality: A Guide for Management. Chapman & Hall. Londres.

Davis, L.E.; Cherns, A.B. (1975): The Quality of Work Life, Vol I y II, New York, Free Press.

Deming, W.E. (1989): Calidad, productividad y competitividad: la salida de la crisis. Díaz de Santos, Madrid

Dueñas, E. (2010): Ciclo de análisis y mejora de las garantías en John Deere Ibérica. Final Project. School of Engineering. University of Carlos III. Madrid.

Duran, A. (2005): Análisis del proceso de elaboración e implantación de instrumentos de Responsabilidad Social Corporativa. PhD. Universidad Carlos III. Madrid.

Eisenhardt, K.M. (1989): Building theories from case study research. Academy of Management Review, Vol 14, n° 4, pp.532-550

Eskildsen, J.K; Kristensen, K.; Juhl, H.J (2000): The causal structure of the EFQM excellence model, in Edgeman, R (Eds), First International Research Conference on Organisational Excellence in the Third Millennium, Estes Park, CO, pp.75-83.

Fortuna, R.M. (1991): El imperativo de la calidad. En Grupo de Consultoría de Mejora de la Calidad de Ernest & Young. Calidad Total. Una guía para directivos de los años 90. Capítulo 1. 3ª Ed. Tecnologías de Gerencia y Producción. Madrid.

Galgano, A. (1993): Calidad Total. Díaz de Santos. Madrid.

Glaser, B.; Strauss, A. (1967): The Discovery of Grounded Theory: Strategies of Qualitative Research. Wiedenfeld & Nicholson. London.

Grijalvo, M.; Prida, B. (2005): La implantación de las normas EN 9100 y el Esquema de Certificación "Other Party" en España. DYNA, nov, pp 37-41.

Grijalvo, M.; Prida, B. (2006): Un enfoque estratégico de la calidad. Estudio de un caso. Alta Dirección, n° 247-248, diciembre, pp. 73-82.

Imai, M. (1991): KAIZEN. La Clave de la Ventaja Competitiva Japonesa. 5ª Ed. CECSA. Méjico

Ishikawa, K. (1988): Práctica de los Círculos de Control de Calidad. Tecnologías de Gerencia y Producción. Madrid.

Ivancevich, J.M.; Lorenci, P.; Skinner, S.J.; Crosby, P.B. (1997): Gestión, Calidad y Competitividad. Irwin. Madrid.

Ivey, M.; Carey, J. (1991): The ecstasy and the agony. Business week, October, 21.

Jackson, B. (2003): Gurús Anglosajones: Verdades y Mentiras. Ariel. Barcelona

Jones, R.; Arndt, G.; Kustin, R. (1997): ISO 9000 among Australian companies: impact of time and reasons for seeking certification on perceptions of benefits received. International Journal of Quality & Reliability Management, Vol. 14, n° 7, pp. 650-660.

Kelada (1999): Reingeniería y Calidad Total. AENOR. Madrid.

Lahera, A. (2004): La participación industrial de los trabajadores en la democracia industrial. Catarata. Madrid

Macías, M.A. (2010): Diseño de una linea de montaje de modelo mixto según metodología lean manufacturing en John Deere Ibérica. Final Project. School of Engineering. University of Carlos III. Madrid.

Martínez, I. (2004): Principales consecuencias de la implantación del Esquema de Certificación Aerospacial. Final Project. School of Engineering. University Carlos III. Leganés. Madrid

Martínez, J.L. (1993): La gestión de calidad total en una empresa española: UBISA. Información Comercial Española, ICE. n° 724, pp. 95-104.

Mcllhattan, R.D. (1991): Sistemas de gestión de costes, JIT y calidad. En Grupo de Consultoría de Mejora de la Calidad de Ernest & Young. Calidad Total. Una guía para directivos de los años 90. Capítulo 13. 3ª Ed. Tecnologías de Gerencia y Producción. Madrid.

Merli, G. (1995): La Calidad Total como herramienta de negocio. Una respuesta estratégica al reto europeo. Díaz de Santos. Madrid

Mizuno, S. (1992): Company-Wide Total Quality Control. 6th Ed. Asian Productivity Organization. Japan.

Molleman, E.; Broekhuis, M. (2001): Sociotechnical systems: towards an organizational learning approach. Journal of Engineering and technology Management, n° 18, pp 271-294

Moreno-Luzon, M.D.; Peris, F.J.; González, T. (2001): Gestión de la calidad y diseño de organizaciones. Teoría y estudio de casos. Prentice Hall. Madrid

Muthler, D.L.; Lytle, L.N. (1991): Necesidades de formación en calidad. En Grupo de Consultoría de Mejora de la Calidad de Ernest & Young. Calidad Total. Una guía para directivos de los años 90. Capítulo 7. 3ª Ed. Tecnologías de Gerencia y Producción. Madrid.

Nanda, V. (2005): ISO 9001:2000. Lograr la conformidad y la mejora continua en empresas de desarrollo de software. AENOR. Madrid.

Nieto, C.N. y Ros, L. (2006). Comparación entre los Modelos de Gestión de la Calidad Total: EFQM, Gerencial de Dwming, Iberoamericano para la Excelencia y Malcom Baldrige. Situación frente a la ISO 900. X Congreso de Ingeniería y Organización. Valencia

Prida, B. (2005): Calidad en la universidad. Qualitas Hodie. Excelencia, desarrollo sostenible e innovación, n° 109, pp. 62-67.

Prida, B.; Grijalvo, M. (2007): Un caso real de implantación de "lean manufacturing". Metodología y reflexiones sobre el proceso de implantación. International Conference on Industrial Engineering and Industrial Management. Madrid.

Prida, B.; Grijalvo, M. (2008): The socio-technical approach to work organization. An essential element in Quality Management Systems. Total Quality Management & Bussiness Excellence, Vol 19, n° 4, pp. 343-352.

Rotger, J.J.; Canela, M.A. (1996): Gestión de la calidad: una visión práctica. Beta. Barcelona.

Senge, P. M. (1998): La Quinta disciplina en la práctica. Como construir una organización inteligente. Granica. Barcelona.

Senge, P. M. (1995): La Quinta disciplina. Como impulsar el aprendizaje en una organización inteligente. Granica. 3ª Ed. Barcelona.

Suzaki, K. (2000): Competitividad en fabricación: técnicas para la mejora continua. TGP Hoshin. Madrid.

Tari, J.J. (2001): Aspectos que garantizan el éxito de un sistema de calidad. Forum Calidad, n° 127, pp. 34-38.

Tillery, K.R.; Rutledge, A.L. (1991): Quality-strategy and quality management connections. International Journal of Quality and Reliability Management, Vol 8, n° 1, pp. 71-77.

Trist, E. (1981): The evolution of socio-technical systems. Quality of Working Life Centre. Toronto Ontario.

Walton, M. (1986): The Deming Management Method. Putnam, New York.

Yin, R.K. (1989): Case Study Research: Design and Methods. 2ond Ed. Sage Publication. London

Zink, K.J.; Schmidt, A. (1998): Practice and implementation of self-assessment. International Journal of Quality Science, Vol. 3, nº 2, pp. 147-170.

12

Using Total Quality Management Model to Face the Economic Crisis: The Case of Mercadona

Miguel Blanco Callejo
Universidad Rey Juan Carlos
Spain

1. Introduction

The aim of this chapter is to describe the recent evolution of Mercadona, a leading Spanish family-owned supermarket company, that has become a worldwide benchmark in retailing industry (Blanco and Gutierrez, 2010; Ton and Harrow, 2010). The chapter presents the company and describes the 1993's top management decision of implement Total Quality Management (TQM) as well as the outstanding and rapid growth of the firm between 1995 and 2008 under that management model (Blanco and Gutierrez, 2010). The surge of the economic crisis which has its origin in the United States in the summer of 2007 started to affect significantly Spain in the second half of 2008. The crisis has had a dramatic impact in the Spanish Economy and Society declining the GDP -3.9% in 2009 and rocketing unemployment rate to 21.5% in October 2011 (INE, 2011). In September 2008 consumer spending dropped off sharply affecting Mercadona's turnover. This situation was a real challenge for the top management team and motivated a significant realignment of the company according to its management model goal: "anticipating and meeting its customers' needs". Anyway, the strength of Total Quality Management has allowed the company to face successfully this situation by refocusing in its management model and adopting innovative and pioneering initiatives in the Spanish retailing industry.

As a supermarket chain providing basic food and hygiene products to its customers Mercadona started to suffer a sustainable client's loss in the second half of 2008. Facing the situation the company has put the customer back at the center of the company's decisions, taking only those steps that provide customer value (Mercadona, 2010; Ton and Harrow, 2010). Given this context Mercadona reviewed all its references, tightened up the product assortment and has launched a campaign to reduce its prices as much as possible "pinching pennies in all its processes" in order to provide the "shopping cart menu" to its customers, offering the highest quality products at the market's most economical prices. Mercadona's recent economic and financial results seem to show the adequacy of the firm's strategy to overcome the economic crisis and to continue firmly with its growth and outstanding development in the upcoming future.

2. Mercadona

Mercadona is a Spanish family-owned supermarket company dedicated to the commercial distribution of food and hygiene products. The firm runs large supermarket (averaging between 1,300 and 1,500 square meters of retail space) following a model of local city centre stores. The 63,500 employees' company finished 2010 with 1,310 urban supermarkets representing a market share of 13.1% of the total food store retail space in Spain (Mercadona, 2010). Mercadona's objective is to fully satisfy the grocery, home cleaning, hygiene and pet product needs of its customers and their pets. The company would like to be a "prescriber" that means that Mercadona recommends to its customers the best products. Every one of the supermarkets has an ample and efficient selection that includes up to 8,000 different items which combine Mercadona's brands and other commercial brands highly recognized by the customers. This large selection of products pretends to satisfy necessities of the 4.4 million households that usually place their trust in the company every year (Mercadona, 2010).

The company name Mercadona S.A. (Initial letters from Spanish Sociedad Anónima, public-limited company), appeared in 1977, but the company actually originated in *Cárnicas Roig*, a family firm owned by Francisco Roig Ballester specialized in carving and selling meat. The firm subsequently evolved into a chain of eight grocer's shops (Mercadona, 2010). In 1981, one of Francisco's sons, Juan Roig, now running the firm, along with his brother Fernando and sisters Trinidad and Amparo bought Mercadona from his parents. At that time, they had eight stores with some 300 square meters of retail space each. It was a small commercial chain of various stores, which soon began to growth in size, first in the Valencia region and then in other areas of Spain. The firm's expansion was coincident with a period of growth in the commercial distribution sector in Spain, which had various causes, in particular, the increasing concentration of the population within cities and the massive entry of women into the labor market. This growth attracted the attention of the large European distribution chains, which began to enter in the country. The strategy of these large companies was to open major retail outlets (hypermarkets), and offer low prices in particular products accompanied by strong promotions, special offers and discounts that were supported by aggressive advertising campaigns in the media (Navarro, 2005). The firms achieved low prices by putting pressure on their suppliers to cut their prices as far as possible. Customers attracted by the low prices on some products also bought other products not on special offer that yielded high profit margins (Blanco and Gutierrez, 2010).

As a consequence of this strategy, in the 1990s the food distribution sector became highly concentrated and foreign multinationals controlled a large part of the market. The competition from these firms intensified the rivalry in the industry, which was now mature, with strong competition and low margins. In 1990, Juan Roig, along with his wife, Hortensia Herrero took over control of the company. After that, in the early 1990s, Roig faced the first great challenge for the firm. As a consequence of the arrival of the International Retail chains profit margins decreased significantly all through the supermarket industry. Mercadona responded to this highly changeable and turbulent environment by adopting a similar policy to that of the large hypermarket chains. The retailer put downward pressure on its suppliers' prices and ran intensive advertising campaigns to promote its special offers every day. But the results were not as good as expected, since this was the same strategy as the big hypermarket chains'. Mercadona was selling more, but every year its profits declined[1].

---

[1] In 1990, with a turnover of €763m, Mercadona's profits were just under €15m. A year later its turnover rose to €877m, but its profits fell to under €6m (Navarro, 2005).

This was a very tricky time for Mercadona, which had to face some extremely difficult situations. Nevertheless, the President ignored tempting offers to buy the company and stuck firmly to his commitment to remain in the business. Thus, in 1993 Juan Roig decided to implement a strategy that broke radically with the dynamic of the sector: he introduced the TQM model as the basis of all its operations. This involved a strategic switch from the Supermarket industry's traditional high-low pricing with promotions to a new commercial strategy summarized in a simple slogan "Siempre Precios Bajos" (SPB, which translates as "Always Low Prices") and a culture of quality and continuous improvement (Blanco and Gutierrez, 2010; Ton and Harrow, 2010).

While its competitors continued with their massive daily advertising campaigns in the media, Mercadona immediately stopped all its advertising spending. Mercadona cancelled all its special offers, and made a commitment to its customers always to sell at the same price, and to its suppliers to remain loyal to them and maintain stable prices for years (Caparrós and Biot, 2006). This was quite a turnaround, as suppliers had previously seen Mercadona as one of the toughest negotiators. The firm's main objective was to protect itself in the midst of all this turbulence and implement an unusual and original model, novel in the commercial distribution sector: price, supplier and employee stability to achieve permanent customers. The notion was to allow the customers enjoy a Total Shopping with products of the highest quality at the lowest prices on the market keeping the motto "quality doesn't have to be more expensive" (Mercadona, 2010)

The results immediately after adopting the new model were not very promising: although the firm managed to practically double its sales from four years before, its profits were less than half. But the President stuck by his decision and firmly maintained the strategy based

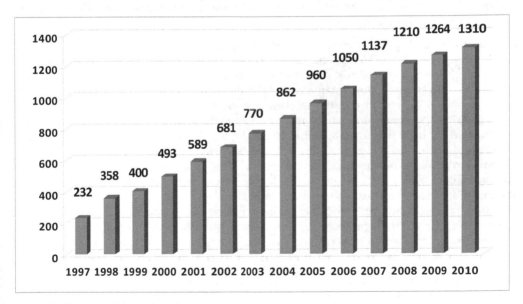

Fig. 1. Evolution of Mercadona's stores
Source: Mercadona's Annual Reports

on the new management model. In 1995 its performance improved, and Mercadona soon entered a spectacular and unstoppable process of growth, which was mainly organic. From 1995, the focus on TQM worked well for Mercadona and between 1998 and 2008, Mercadona enjoyed double-digit growth in sales and profits. Supermarkets spread out throughout the whole country like an olive-oil spill. By the end of 2008, it was the biggest supermarket chain in Spain, with 1,210 stores and 61,500 employees (Ton and Harrow, 2010). Moreover, in recent years Mercadona has maintained a rate of new supermarket openings of nearly 100 per year (Figure 1). The growth rate in its sales of over 25% per year from 2000 to 2006 makes Mercadona the 14th fastest-growing retailer in the world, and in 2006 it was the fastest-growing food distributor in the world, after the American commercial distribution giant Wal-Mart, (Deloitte, 2007).

3. Mercadona's total quality management model

Mercadona's management firmly believes that adopting and maintaining the TQM model has been critical in allowing the firm to achieve this rate of growth and these spectacular economic results. As Juan Roig, Mercadona's president stated: "Without doubt one of the greatest milestones of Mercadona was the implementation of the Total Quality Model. The involvement of all those whom we train at Mercadona in applying and developing it is doubtlessly the reason for our success" (Ton and Harrow, 2010: 2). Through this approach, management has impressed on the firm a clear orientation to satisfy the needs and expectations of all its stakeholders. On the basis of this approach Mercadona has defined five components in the firm: first, customers ("the Bosses"); second, the employees; third, the suppliers; fourth, society, and fifth, capital. All are equally important, but in that sequential order, (Mercadona, 2010). As the president of Mercadona noted "Mercadona's mission is not just to make a profit, but to make it while satisfying all of the five components" (Ton and Harrow, 2010)

Mercadona's TQM model starts from a universal premise: "to be satisfied, first you have to satisfy everyone else", (Mercadona, 2007). Thus, in the first place, Mercadona orients its entire business model towards the complete satisfaction of its customers, which the firm considers so important that internally it calls them its "bosses". In the same sense, the management is convinced that the workers – who are the people who have to satisfy the customers in the stores – must, in turn, be satisfied as well. This is why the firm, has established its own way of managing its employees, implementing policies that seek to encourage their self-realization through job security and stability, training, internal promotion and improvements in their quality of life. It is the suppliers' products that have to maximally satisfy the firm's customers, so Mercadona has established a relationship with its suppliers based on trust, cooperation, mutual collaboration and stability. Society is the medium in which the firm carries out its activities, so Mercadona feels involved with and ethically committed to its protection and development. Finally, by satisfying the needs of its customers, employees, suppliers, and society, Mercadona contributes to satisfying the expectations of the fifth component of its model, capital (Figure 2).

Despite its apparent simplicity, applying the TQM model is neither simple nor easy. In fact, one of the main difficulties is to introduce and develop programs and tools allowing the firm to satisfy each and every one of its diverse components, and at the same time, ensure that the components perceive that the firm is satisfying their needs so the firm transmits the image of a responsible and committed company. In any case, with this powerful model Mercadona has gained competitive advantages over its rivals, has been able to globally orient its strategy and organizational culture and has had adopted a coherent decision-making tool. So, any initiative proposed in the company must first be evaluated for its compliance with the model. The firm adopts and implements proposals only when their consequences are satisfactory to all five components, otherwise it rejects them.

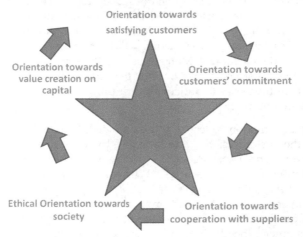

Orientation towards satisfying customers

Orientation towards value creation on capital

Orientation towards customers' commitment

Ethical Orientation towards society

Orientation towards cooperation with suppliers

Fig. 2. Mercadona's Total Quality Management Model
Source: Blanco and Gutierrez (2010)

3.1 Orientation towards bosses' (customers') satisfaction

One of the fundamental cornerstones of the Total Quality Management Model is its orientation towards satisfying customer needs. This is why Mercadona puts the customer in a privileged position and carries out a large number of activities and initiatives to be able to care for its customers, strengthen their loyalty, and help them generate the maximum possible value from doing their shopping in Mercadona stores.

For Mercadona, the customers are at the center of its activity, since its very survival depends on their decision to shop in Mercadona supermarkets or elsewhere. Nevertheless the staff of Mercadona never spoke about their customers; they spoke of their "Bosses". Thus, the firm calls anyone who shops in its stores "boss". Under this premise, the whole organization is focused on providing an excellent service to its customers. This conception places the customer at the top of the firm's inverted organizational pyramid, and the function of the leader and the rest of the organization is to serve the customer. So, everyone's job is to provide "the Bosses" with the best quality-to-price ratio, the correct assortment of products and the highest level of service while ensuring that the Bosses could complete their purchases as quickly as possible (Ton and Harrow, 2010).

The culture of serving the customer prevailed throughout the business. Given the customer's privileged position, Mercadona's philosophy is to continually increase the "value for the customer" by maintaining its "Always Low Prices" policy and eliminating special offers, promotions and temporary discounts. A key service to "The Bosses" is keeping prices not only as low as possible but also as stable as possible (Ton and Harrow, 2010) But the philosophy not only focuses on prices. All the firm's actions are oriented toward satisfying all the customer's specifications, whether known or potential future expectations. In order to find these out, Mercadona maintains a constant and direct dialog with its "bosses", carrying out activities to collect information, opinions and needs from its customers: meetings with neighbors in neighborhoods where new stores are being opened, monographic courses on product lines, open days, blind tests, free customer service hotline, and suggestion boxes in supermarkets[2]. These operations require considerable economic investment from Mercadona, but the firm considers this form of contact more appropriate than big advertising campaigns in the media, which are extremely expensive and collect little information from the customers. Mercadona directly learns about its customers' tastes, the products habitual customers need, and the future trends in different lines of consumption. The firm then tries to anticipate their future needs through innovation, R&D and product improvement programs.

On the basis of the information collected and distributed, Mercadona carries out actions to increase value for the customers. The aim is to give them the chance to carry out all their shopping under the principles of "maximum quality, maximum range, maximum service, minimum cost and minimum time – always". With regard to quality, the retailer stresses the nutritional quality of its more than 8,000 products[3]. But quality also involves convenience, speed, comprehensiveness of service, and product variety. Product variety – the maximum range – does not mean offering a large number of brands, but rather all the products that satisfy all the customer's needs in food, hygiene, cleaning and pet food. Moreover, Mercadona aims to become a "prescriber": choosing and recommending products for their quality and minimum price. Mercadona guarantees the life and name principles for those products, guaranteeing the origin and date of packaging of the products, as well as the same name on products of the same quality. For this, the firm has designed a relationship system with the suppliers of its own-brand products[4]. This reduces customer insecurity, strengthens customers' loyalty to recommended products, and generates trust. With regard to service quality, Mercadona makes shopping as convenient as possible: payment by credit card, loyalty card, online or telephone shopping, home delivery, and free parking for customers. Moreover, the firm improves the convenience of the store by using a very functional design,

---

[2] In 2006, Mercadona used direct communication in 180,000 encounters. This type of action enabled the firm to directly reach more than 1.9m households. The company has 250 employees (monitors) that are collecting information in the supermarkets directly from the clients. In addition to that Mercadona has a customer-service department consisting of more than 90 employees, and the firm responded to 460,000 customer consultations in 2010. Each complaint is perceived as an "opportunity", and the employees are obliged to respond to the customers, Mercadona (2010).

[3] Mercadona's quality department works with its suppliers to guarantee the nutritional quality of all the firm's food products. In recent years, the firm has introduced more fibre-rich foods and green products, substituted saturated fat with sunflower oil, introduced low-salt products, started food traceability programs with its suppliers, and signed collaboration agreements with certification specialists and the Spanish Ministry of Health and Consumer Affairs to encourage healthy consumption habits and improve the population's nutrition, Mercadona (2000-2007).

[4] Mercadona's own brands are Hacendado (food), Bosque Verde (cleaning products), Deliplus (hygiene products), and Compy (pet food). These products are endorsed by both Mercadona and the "intersuppliers", and their quality, traceability and food safety is guaranteed at the lowest possible price (Mercadona, 2010).

what Mercadona calls "Ambience Stores" (Blanco and Gutierrez, 2010). From 2000, the stores are divided into six ambiance store sections (butcher, fish, bakery, fruit and vegetables, cosmetics and deli), each with its own décor and adjusted and designed depending on its marketing requirements. The stores have air-conditioning, shelves are restocked outside opening hours, products come in various packaging formats according to customers' needs, and the firm aims to minimize the time customers spend shopping. The notion behind this is to "provide value to the Bosses' shopping experience". The "Ambience stores" offer Mercadona's customers larger supermarkets with more pleasant atmospheres and a more logical product arrangement (Mercadona, 2010)

The result of this policy was a spectacular growth in the firm's sales: turnover has increased more than nine-fold in the past decade from 1997 to 2007 his growth is not only down to new store openings: the company has also increased its same-store sales considerably increasing it by an annual average of around 8% between 2005 and 2007 (Figure 3).

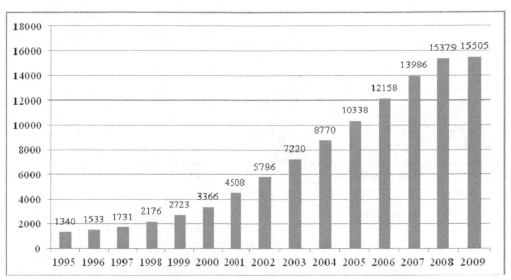

Fig. 3. Evolution Mercadona sales 1995-2009 (€000s)
Source: The Authors, based on Mercadona Annual Reports

3.2 Orientation towards employee's commitment

For Mercadona, the orientation toward the employees is another cornerstone of the TQM model. The premise is that quality is made by people, and all 63,500 employees (67% women) need to be conscious of quality. The firm needs to achieve the commitment, involvement and/or participation of its workers by incorporating certain values that translate into concrete practices in human resources management. So Mercadona's human resources policy is based on providing its employees with stability, good working conditions, training, transparency, compensation, quality of life and opportunities for advancement (Ton and Harrow, 2010).

Within Mercadona's management, Total Quality is based on the universal premise that "in order to be satisfied, one must first satisfy others". The search for the total satisfaction of what Mercadona considers its most valuable asset, its employees is "relentless". To this end, Mercadona's human resource model is based on leadership and relies on self-realization, permanent employment, ongoing training and striking a balance between work and family life for everyone at the company. This makes it possible for the person who has to satisfy "The Boss" to in turn be satisfied (Mercadona, 2010)

With regard to staff selection and recruitment, Mercadona looks for people who fit in a quality environment. Mercadona tried to recruit store employees who would fit its carefully developed and distinctive culture. Candidates must be at least secondary-school graduates, and the selection process involves cultural knowledge and psycho-technical tests, interviews, and group dynamics. Recruiters held meetings at which they interviewed applicants, explained the job conditions, and then placed the applicants into groups where they develop a role-playing scenario. Recruiters received close to 30 applicants per opening and typically interviewed three for each position (Ton and Harrow, 2010)

Employees' training begins as soon as they join the firm with the so-called "Welcome Program". Before they start work, all employees receive 4-weeks, full-time training on the TQM model, which aims to inculcate the firm's culture and cost 5,000€ (Ton and Harrow, 2010). To aid learning the firm gives all its workers manuals outlining the main ideas of the model, plus practical exercises that are subsequently evaluated in exams and written tests. After taking up their posts, the workers receive very varied training during their time in the firm, including courses on how to use new technologies, task specialization in the supermarket, leadership, and occupational safety and health[5]. The training programs are taught by staff members and evaluated internally. Moreover, they often take place in the supermarkets themselves, all of which have specially-designed training rooms. The aim of the training policy is to develop employees' skills, and the policy is accompanied by an internal promotion policy that facilitates employees' identification with Mercadona. In fact, all the firm's senior managers started at the bottom.

Mercadona has designed a simple and complete compensation policy aimed at strengthening its link with its workers and favoring their identification with the project. Part of each employee's compensation is fixed, and the other part is variable, satisfying the equity principle: "for equal responsibility, equal pay". The fixed component has only four bands, and is based on three criteria: experience, responsibility, and performance. The variable component depends on the company and each individual supermarket meeting their objectives. In any case, Mercadona pays wages above the sector average, and the firm consequently demands more commitment from its workers in exchange. The evaluation of the performance and quality that the employee shows in their work, which is carried out by the employee's immediate superior, is determinant in pay rises and promotions. It is significant that the majority of Mercadona's employees earn bonuses for meeting objectives[6].

On the other hand, jobs are designed broadly, which enhances both the firm's flexibility and the workers' employability. Each worker knows half an hour in advance what their function

---

[5] Training investment amounted to €30m in 2010 an average of 471€ per employee. The firm fave more than 1,2m hours of training in 2010 (Mercadona, 2010).

[6] Bonuses totalled €25m into 2001, €52m in 2002, €64m in 2003, €81m in 2004, and €104m in 2005. In 2006 Mercadona had foreseen a bonus of €124m, but its net profits of €242m meant the firm distributed an extra €43m. Each worker earned a bonus of between 1,500 and €3,000, Mercadona (2001-2006). In 2010 in the midst of global financial crisis bonuses totalled €210 (Mercadona, 2010)

will be that day, and this depends on variables such as the number of customers, restocking needs, and staff available in the store. The firm has tried to standardize different types of timetable through its "standard timetable" program. All timetables have the same start and finish times regardless of the day of the week, and staff members have a maximum number of hours work per week. Mercadona workers also have a concentrated working day. In other words, they do not work the split shifts — with a long lunch break from 2 to 5 o'clock — that are typical in Spanish stores. This program also allows workers to know their timetables one month in advance and hence be able to plan their 30 days per year vacations (Mercadona, 2010).

Mercadona also aims to reconcile its employees' working and personal lives as far as possible through stability programs, additional services and benefits, as well as protect its workers to the maximum. In a sector where temporary contracts are the norm, all of Mercadona's staff has stable, indefinite-duration contracts. Moreover, Mercadona continually tries to improve its employees' working conditions. The firm offers free kindergartens in its logistics centers, pays for one month extra maternity leave, does not open its supermarkets on Sundays, and has a policy allowing workers to work in the supermarket closest to their homes. Finally, Mercadona also carries out preventive and corrective activities in the area of occupational safety and health through a culture of internal prevention that measures and actively combats accidents and illnesses. In addition, all the firm's employees have life insurance. Mercadona pays employees 100% of their wages if they are incapacitated, and the firm has measures in place to help the family if a company worker dies.

With all these actions included in a comprehensive human resources policy and in the framework of its TQM model, Mercadona has managed to considerably reduce its employee

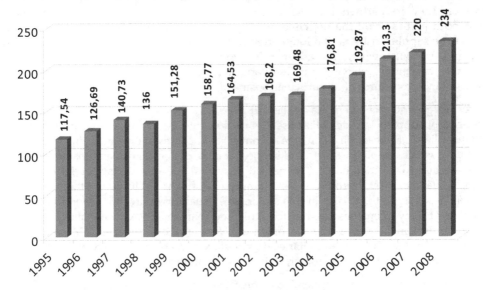

Fig. 4. Evolution Sales/Employees ratio 1995-2008 (€000s)
Source: The Authors, based on Mercadona Annual Reports.

turnover and absenteeism figures, which are traditionally very high in the sector. At the same time, the firm has also raised its workers' productivity levels considerably, and consequently its own results. The following figure shows the upward trend in sales per employee (Figure 4).

3.3 Orientation towards cooperation with suppliers

Incorporating the principle of external cooperation into the TQM value system involves extending cooperation beyond company limits. Mercadona puts this into practice with its suppliers in a special way. Mercadona's relationship with its suppliers seeks cooperation, stability in the relationship, mutual benefits, and the sharing of training and information to improve all the business processes.

The starting point is the adoption of a model of "process management and analysis". Mercadona studies and analyzes these processes from when the products leave the supplier to when they reach the customer. The central idea is that any inefficiency in the production process will eventually reach the end-customer. The aim is to iron out inefficiencies, and to try to extend cooperation to the whole value chain and even reach the suppliers' suppliers.

Mercadona believes that in order to optimize process management it is necessary to generate a relationship of mutual trust with its suppliers. This relationship starts from the premise that the continuity of the supplier-Mercadona relationship is in the hands of the end-customer. The supplier will continue to supply Mercadona if it can respond to and satisfy customer needs. Thus, the supplier becomes a fundamental element for satisfying the customer, and according to Mercadona's philosophy, the supplier must also be satisfied with its relationship with Mercadona, generating synergies between both parties. Mercadona's suppliers share certain generic characteristics: they are partners, capitalists, and focused on adding value for the customer. They are partners because the firm establishes a relationship with Mercadona in which both parties share the objective of fully satisfying the customer. They are capitalists because the supplier risks capital in an attempt to earn profits. And what is most important, the firm must be focused on the customer. It must be able to offer value thanks to its expertise in its area of activity: carrying out R&D and innovation, and knowing the raw materials, processes and procedures of its particular business.

Given the generic characteristics of its suppliers, Mercadona distinguishes between four categories of supplier: classic, "distressed" suppliers, intermediaries, and what the firm calls "intersuppliers". Mercadona maintains a conventional contractual relationship with "classic suppliers", which supply products and services that the customers demand mainly as a result of these companies' marketing. "Distressed suppliers" are producers that have difficulty finding a market for their products. Mercadona acquires their products, and these firms' very survival depends on their sales in this supermarket chain. Working with Mercadona allows these firms to stay in business, and hence avoid exits that would have traumatic effects on the workforce. Intermediaries are agents operating between the supplier and Mercadona. The retailer believes that these intermediaries do not add value, are unnecessary and only increase the cost of the product to the customer, so it tries to avoid them.

Finally, the so-called "intersuppliers" or "integrated suppliers" (Ton and Harrow, 2010) are Mercadona's fourth category of supplier. Mercadona works with over 2,000 suppliers, of whom approximately 100 were "integrated suppliers" who supplied it home-brand products (Mercadona, 2010; Ton and Harrow, 2010). The Integrated Suppliers, manufacturers of Mercadona's brands and with which the company has a long-term mutual cooperation and commitment relationship regulated by way of a Sound Business Practices Framework Agreement and which results in "contracts for life" (Mercadona, 2010).

These firms share Mercadona's TQM-based management philosophy, and their relationship with Mercadona is long-term and indefinite, with a willingness to undertake activities jointly. Mercadona's management is committed to the growth, development and sustainability of its "intersuppliers", and the firm dedicates substantial resources to joint research and product improvement, which help improve these firms' competitiveness and their chances of satisfying customer needs. The intersuppliers products exclusively for Mercadona and take on the "Totaler" philosophy, that is, to be prescribers of products that satisfy all the needs of the customers – the "bosses". They should offer customers recommended products that figure among Mercadona's store brands that have the highest possible quality at the lowest possible price. These suppliers must satisfy some requirements in order to obtain the "intersuppliers" status. Firstly, the company owners must have "passion" for what they are doing, that is, they have to be proactive, innovative and completely involved in their business. They must also have the economic resources necessary to carry out their business goals as well as the management ability to develop them. Secondly Mercadona sought companies with stability on the stock market, as they transmit security to the relationship. Furthermore, they must be companies that have an open-minded attitude with respect to information and indicators coming from customers and the market in order to provide solutions to meet customer's demands. The candidates must also be willing to introduce the TQM model in their organizations and seek to satisfy all of its components, which include the study and modification of every aspect with the aim of improvement, guaranteeing nutritional and environmental safety, and willingness to be audited by Mercadona to verify the degree to which they fulfill the model's requirements. Finally, Mercadona requires the exclusive manufacturing of some products as a result of the relationship of mutual trust and cooperation between both companies.

The process by which a firm becomes one of Mercadona's "intersuppliers" or "integrated suppliers", which is one of the most innovative and original aspects that Mercadona has contributed to the TQM model, is long and complex, and consists of three stages. In the preliminary or approach stage, candidate firms must occupy a position of leadership and have a consolidated and reliable production system. In the second stage, the firm and Mercadona establish a relationship where they share all information—a true partnership. In the third stage, the supplier must produce in response to customer needs and starting from their expectations and demands. This is why Mercadona shares the information it obtains from its customers through various mechanisms with its suppliers, and works with them in projects to produce new products.

This is when the relationship fits perfectly with Mercadona's management philosophy, which is to try to become a "prescriber" of products for all its customer's shopping. For this, Mercadona must use the information obtained to inform its "intersuppliers" of customer needs, and the two firms must jointly develop products and services, apply the TQM model in a coordinated way, and use the value-effort model to measure each other's value generation.

3.4 Orientation towards society

Analyzing the impact of the firm's actions on its social environment is also a defining aspect of firms with a management model based on total quality. In fact, it is a principle of TQM that, by contributing to the development of society, Mercadona would increase its own potential for development. The company attempted to detect what society needed through its communication with customers, as well as through business organizations and municipalities (Ton and Harrow, 2010). Mercadona's social strategy runs a large number of initiatives seeking social advances in various aspects: education and research; improvement and care for the natural environment; programs to help people with health problems and disadvantaged people; revitalization of trade in urban areas; participation in sectorial associations of commercial distribution; and finally, creation of wealth and jobs in the Spanish agro-food sector.

In education, Mercadona participates in initiatives linking university studies with business activity, fundamentally in the region of Valencia, and runs research programs in the agro-food field. To care for and improve the natural environment, the firm has adopted the principle of behaving responsibly toward the environment incorporating "environmental productivity" (Ton and Harrow, 2010). The firm runs waste recycling programs, studies waste treatment, implements measures and global solutions aiming to minimize the environmental impact of its activities, seeks mechanisms that achieve maximum energy efficiency, and optimizes merchandise distribution systems both nationally and in the cities[7].

With regard to Mercadona's assistance for social groups with health problems and disadvantaged people, the supermarket chain sells special gluten-free products for customers with celiac disease, and sugar-free products for diabetics, thereby helping these groups with their shopping. The firm has also signed agreements with public institutions to promote joint measures aimed at helping victims of domestic violence to find employment, and professional and social training programs for young people with social problems and difficulty entering the labor market. Mercadona has also contributed visibly to revitalizing and re-energizing local city-centre trade, for this, the firm has restored and re-launched traditional Spanish indoor markets situated in the city centre, putting a supermarket inside.

Mercadona participates in sectorial and wealth-generating associations such as the Spanish Association of Supermarkets (ASEDAS) and the Spanish Association of Commercial Codification (AECOC), which work in favor of the development of the commercial distribution sector. With regard to wealth creation, it is interesting to note the

---

[7] Mercadona has a commitment with sutainable Transport under the principle of "Transport more with fewer resources". The company is using a combination of trucks, trains and ships to reduce the environmental impact. With regard to the logistics system in cities, the firm has implemented the Urban Merchandise Transport system (TUM) with Silent Nightime unloading in more than 407 stores. In this system, products are unloaded to supermarkets in the early morning from large, silent articulated lorries. This system cuts noise pollution, traffic congestion and the negative impact on the environment: the reduction in CO_2 emissions is equivalent to 30,000 cars. Meradona invested 23 in the environment in 2010. In addition to that Mercadona was recognized by the European Union as an example of a company that fosters good environmental practices (Mercadona, 2010).

retailer's important and growing contribution to Spain's GDP[8] its leadership in the generation of stable employment in Spain, and its substantial investment in the Spanish agro-food industry.

Finally, the most notable of Mercadona's characteristics is that the company is not content with just implementing initiatives that show its commitment and responsibility to society. The firm also strives to measure the impact of these policies. With this purpose, in 2002 Mercadona voluntarily commissioned an ethics audit from an independent association. This resulted in a highly satisfactory assessment of the firm's compliance with the requisite ethical standards, and according to the management this should serve as reference for the development of the firm's corporate social responsibility policy[9]. Moreover, in another external study from an independent, international organism—the report on corporate reputation that the Reputation Institute publishes annually—Mercadona has the fourth best corporate reputation in the world, behind only Lego, Ikea and Barilla[10] (Reputation Institute, 2007).

3.5 Orientation towards capital

The company's TQM model also has repercussions at the level of economic-financial results and value creation for its owners. Mercadona is a family firm that is not listed on the Stock Exchange. Its stock ownership is highly concentrated: the majority of the ownership and control is in the hands of the company's President, Juan Roig, and his family.

The President argues that under the TQM model it is also "fair" to satisfy the needs of the company's final component, the capital, in other words, those people and organizations that invest their money in the firm and hence provide economic resources. This means that Mercadona also has a clear orientation toward profit maximization, as a means of satisfying its stockholders' needs and desires (Caparrós & Biot, 2006). The company's management model has the aim of offering its stockholders various advantages that they will regard as valuable, such as profitability, stability, security and minimum risk in their investment. Mercadona was focused on long –term profitability, continuous improvement in productivity and reinvestment of profits to increase equity. In line with this position, the firm's economic performance can be classed as spectacular. The company's net profits have increased more than 14-fold from 1997 to 2007, reaching €336m in 2007. Figure 6 in the next page shows the evolution in both gross and net profits after the company introduced its TQM model.

---

[8] Mercadona's contribution to Spanish GDP was €960m in 2002, €1,233m in 2003, €1,471m in 2004, €1,756m in 2005, €2,140m in 2006, €2,516m in 2007, €2,681m in 2008, €2,662 in 2009 and €3,059m in 2010,(Mercadona, 2010).

[9] The ethics audit, commissioned from the Spanish Foundation for Business Ethics (ETNOR), aimed to provide a reliable picture of the perception that the five components have about the firm. ETNOR carried out 2,000 contacts consisting of personal interviews and telephone questionnaires. Respondents were asked about how they felt they had been treated in relation to the following aspects: integrity, credibility, fairness, dialog, transparency, dignity, legality, corporate citizenship, environment, and responsibility. Mercadona was awarded 4.31 points out of 5 for its ethical standards. The firm committed itself to repeat the audit every three years, ETNOR (2003), Mercadona (2003-2007).

[10] In 2010 the company held the tenth place in the Reputation Institute ranking and it has become one of Spain's two leading private companies in terms of its reputation, according to an annual study conducted by KPMG (Mercadona, 2010).

With regard to the distribution of its profits, the firm re-invests a large part of them, thereby favoring firm growth. This has allowed the firm to grow organically and carry out a spectacular expansion strategy, opening new supermarkets, warehouses and logistics centers throughout the whole country in recent years. Figure 5 shows the evolution in the firm's equity and investments. This is the result of a careful policy of profit distribution allowing the firm to strengthen and increase its equity and multiply the investments that enable growth and expansion. This has ensured that the TQM model has had positive and multiplicative effects.

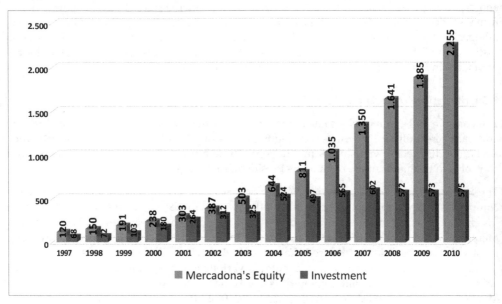

Fig. 5. Evolution in Mercadona's equity and investments (€m)
Source: The Authors, based on Mercadona Annual Reports.

The evolution in the firm's main economic indicators clearly reflects, in the words of the management, the "idealness" of the Total Quality Management model (Caparrós & Biot, 2006), and is the result of satisfying the other four components of the model: customers, employees, suppliers, and society. Continuing the current policy, as well as improving and consolidating the TQM model, should ensure the future viability of the project, confirm the trend in improved results, and increase the value of stockholders' capital.

4. Facing the economic crisis: The strength of Mercadona's total quality management model

The global economic crisis, which began in September 2007 with the bursting of the U.S. property bubble has seriously affected Spain by late 2008 (Ton and Harrow, 2010). During the second half of 2008 the effects of the Global Economic crisis started to be clearly noticed in Spain. The crisis has been characterized in Spain by a high rate of unemployment, significantly above the average European rate and an important credit restriction in capital markets (Alonso and Furio, 2010). Actually, Spain has recorded the highest rate of unemployment in the European Union lately. This situation logically had a considerable impact in consumers' buying decision who tried to adjust as much as possible their budgets as well as reduced the number of transactions.

Facing such a situation, Mercadona's top management team was aware from the data from the reports tracking market's evolution that the company was starting to loss customers in its stores. According to its 2008 reports, the clients were starting to consider Mercadona "an expensive supermarket". This situation was translated into a significant reduction in Mercadona's profit that broke with an increasing trend from previous years (Figure 6)

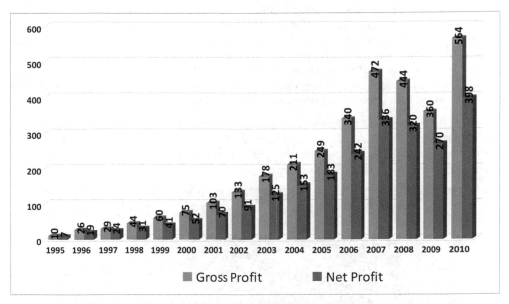

Fig. 6. Evolution in Mercadona's gross/net profits 1995-2006 (€m)
Source: The Authors, based on Mercadona Annual Reports.

Mercadona's Top Management team started to be really worried with business and profits evolution and tendency and it decided to undertake a deep strategic reflection about its business. As a result the company was able to justify the situation based in internal and external causes. Externally, the company concluded that the situation characterized by an extremely high unemployment rate and clients' difficulty to access credit, economic crisis was significantly damaging Mercadona's business. Internally, Mercadona's president noted "Based on plenty times and the belief that growth was going to last forever we had been sleepy" (Mercadona, 2008).

Taking into account this starting point, Mercadona decided to provide an answer and face that situation putting "The boss" once again at the heart of every decision the company made. Mercadona decided it had to lower its prices. Firstly, the company decided to perform a full review of the assortment of 8,000 references that it was offering to its "bosses" in the stores. Such analysis was made with the clear aim of provide an answer in line with "reduce prices" and using two criteria:

1.   Did the products and commercial formats satisfy customers' needs? (Need)
2.   What is the level of product turnover? (Rotation)

The analysis allowed the company to realize that it had product in the shelves which were not adding value to its customers. In 2008 Mercadona sold, for example, 72 types of milk, 112 types of juice, 100 types of coffee, carrots of every shape and size, and several types of tomato sauce. The products in some categories, such as coffee, varied only slightly (Ton and Harrow, 2010). So the company decided working to do away with anything that does not add value to the process, that raises product prices and that affects directly the grocery budgets of the customers, who "are only willing to pay for real improvements" (Mercadona, 2010)

The review was made taking into account two basic assumptions, first to eliminate everything that was not adding value to "the bosses" and, second, trying to discover everything that improving quality and product safety was, in fact, adding value. As a result of the analysis, the company was conscious that it had numerous products and commercial formats in basic and highly consumed product like milk, fruit juices, coffee and tomato sauces as well as in cleaning and hygiene products that were not satisfying customers' needs and consequently they are not adding value to the clients.

According to the analysis performed, the company decided to eliminate 800 products of the assortment, being half of them "recommended" Mercadona's branded products. The products eliminated were shown to the media in the conference press when the president presented Mercadona's 2008 results (Figure 7).

Fig. 7. Some of the references eliminated by Mercadona's shelves
Source: Mercadona Annual Report (2008)

Once the analysis had been made, the company decided to realign its strategy again with TQM model and under one of the basic assumption of its competitive strategy, "Reduce Prices". Mercadona's top management team focused the strategic review as a return to "Total Quality Management model" and to the core activity of the firm "to be storekeepers".

The idea was to satisfy clients demands and needs with "a cart will all the needed products with the best quality and with the lowest price, prescribing them the best relationship between quality and price for each product and giving them the option of being right in their shopping decisions". Mercadona named that cart "the shopping cart menu" (Mercadona, 2008:7).

The notion of "Shopping cart menu", as a complete assortment of products that allows the company to satisfy all the clients' needs is the answer from a basic commercial learning made by Mercadona. The client was not worried by the individual price of each one of the different products that were in the cart on the contrary he was worried for the final price of the whole cart (Figure 8). Given that basic assumption, if the company is able to reduce the price of the whole shopping cart, it will be perceived as the supermarket chain cheaper in the market, which, in an environment characterized by the economic crisis will attract an increasing number of clients.

Fig. 8. Mercadona's 2008 Annual Report Cover
Source: www.mercadona.es

So, Mercadona designed in 2008 a complete program of actions oriented to reduce the price of the shopping cart for its clients. The program considered a full review of both Mercadonas' products and business processes with the aim of eliminate all the elements that were adding cost without adding value to clients. After the businesses process review and the subsequent implementation of more than 600 hundred innovative initiatives the company saved more than 500 million euro cooperating with its "integrated suppliers". Between the most outstanding initiatives of cost reduction it was a sound action the elimination of trays, fixed weight and plastic in the packing of fruit and selling them loose implied a saving of 0,35 euro per Kilo which along 2009 was translated in an annual total save of 175 million euro (Mercadona, 2009).

As a result of this program and under the principle of "Back to basics" (Volver a la sencillez), Mercadona started to lower the price of the products effecting directly the reduction in the price of the raw materials, going to an efficient assortment and reducing firm profit but increasing the volume of products sold. The idea was that through the lower cost in the businesses processes the company was going to have lower prices and higher sales (Ton and Harrow, 2010). In this context the company is not defining economic objectives but quantity or volume of products sold objectives. In order to do this, it has defined a new measure from the kilo and liter sold the "Kiliters" according to which it has defined its objectives.

The result of this strategy has been a reduction of 10% in the prices in Mercadona's supermarkets along 2009 (Mercadona, 2009) and a 14% drop in prices since 2009 (Mercadona, 2010). According to Mercadona's estimates it has meant a monthly household average spending reduction of 60 euro and of 720 euro per year for its customers (Mercadona, 2009). The search for excellence in each and every one of Mercadona's processes allowed the company to offer its customers savings in excess of 700 million euro in 2010 stemming from a 4% price reduction in the "Shopping Cart Menu". If the company adds the 1,500 euro in savings in 2009 and which continued to pass on in 2010 the total savings was 2,200 million euro in 2010. None of this would have been possible without the proven dedication of the company's integrated suppliers and the commitment of its employees (Mercadona, 2010).

5. Conclusion

This chapter shows that the total quality management approach has oriented Mercadona's management model and helped the firm to achieve spectacular economic results and a profitable and outstanding growth (from 1995 to 2007) as well as face successfully a challenging situation of economic crisis (from 2008 to 2011). The key aspect of the model is the high consideration awarded to its five components, as well as the orientation of the whole company toward satisfying their needs and expectations.

The chapter shows that Mercadona's top management decision to orient the company's entire management towards satisfying customer needs was correct. But the chapter also shows that the company cannot achieve an optimal management of customer needs without also satisfying the needs of the firm's other stakeholders. Thus, Mercadona's human resource management implements initiatives that strengthen employees' commitment to and involvement in the firm's project. The firm has installed a policy of cooperation, trust and mutual collaboration in its supplier management, which ensures that the products on the supermarket shelves match customers' requirements and specifications more closely and could be adapted in case the economic situation required that. Nor does the firm ignore the social and economic context in which it operates. This can be seen both quantitatively in terms of the creation of wealth and value, and qualitatively in terms of its ethical and environmental commitment with Spanish society and citizens. Finally, with regard to the capital, the economic results of the model are spectacular and have catapulted the company to a position of leadership among Spanish supermarket chains and a competitive advantage that its rivals will find difficult to match. In fact its competitors are currently reacting by imitating its management model and copying some of its initiatives and policies.

So, Mercadona offers the lowest prices in Spain, and its operational performance exceeds that of comparable Spanish and foreign chains. In 2008, Mercadona's sales per square foot was 60%

higher than that of France's giant Carrefour, and more than twice that of an average U.S. Supermarket. Sales per employee were 18% higher than that of other Spanish supermarkets that disclosed financial information that year and more than 50% higher than U.S. supermarkets (Hanna, 2008). The difference is Mercadona's Total Quality management model (Table 1), as Mercadona's president noted is "the reason for our success" (Ton and Harrow, 2010:2).

The boss	– Promoting dialogue and direct communications with our "Bosses" – Maximum quality at the lowest price: Recommended Products. – A return to basics so as to reduce prices with a quality, efficient selection. – Shopping Cart Menu: The highest quality and lowest priced Total Shopping on the market. – Closeness and proximity. – Innovation across the board oriented toward their needs (product, technology, concept and processes). – Always Low Prices (SPB): Price stability, without sales or promotions. – Prescription: to inform "The Boss" and recommend those products with the best price to quality ratio on the market.
The employee	– Stability (permanent contract), professional development and promotion from within. – Fixed responsibilities. Employees carry out their functions in the same position and changes are avoided; this facilitates employee specialization so that they can do what they know best: satisfy "The Boss." – Work and Family Life Compatibility by, among other measures, not opening on Sundays. – Human Resources management model based on individual and collective leadership. – Permanent dialogue. – Profit sharing. – Necessary and continuous training for the conduct of their job functions.
The Supplier	– Stability/Contracts for life. – Long-term agreements, communications and transparency in the relationship. – Bolstering productive activity to promote development and wealth. – R&D plus 'double I' (Research and Development plus Innovation, supported by Investment). – Promoting innovation and process optimization so as to offer the product with the best price quality ratio on the market.

The Society	– Commitment to the social and economic development of those areas in which we have an established presence.
	– Social productivity:
	   o Produce more with fewer natural resources.
	   o Sustainable transport.
	– Striving to "be invisible" in the setting where we carry out our activity.
	– Permanent dialogue, active participation and closeness.
	– Training and reporting on the company's Model and on activity and decision planning.

The Capital	– Constant and sustainable growth. – Reinvesting profits. – Innovation as a factor in competitiveness. – Focusing on the long term. – Increasing productivity through reengineering and process standardization. – Building our relationships on commitments

Table 1. Mercadona's Total Quality Management Model
Source: 2010 Mercadona's Annual Report

In any case, the above describes an outstanding example of the successful application of the TQM model in the highly competitive, mature commercial distribution sector. Reading this chapter should encourage managers to consider the implementation of this model in other sectors of activity, although clearly, and as the literature observes, firms will need to make certain adjustments to the actions and policies depending on their particular industry (Wicks, 2001) and geographical area (Lagrosen, 2002). Proof of the feasibility of introducing this model in other industries and sectors comes from the fact that the model has spread sucessfully to some of Mercadona's intersuppliers or "integrated suppliers". Indeed, some of these have achieved even better results and growth rates than the supermarket chain itself (Blanco and Gutierrez, 2010).

Finally, Mercadona's Total Quality management model, thanks both to its feedback mechanisms and its great capacity to adapt to environmental and dramatic economic changes, is guaranteeing the future of the firm, even in a challenging economic situation like the one that Mercadona is currently facing in a country with an unemployment above 20% and a significant difficult to access credit. In spite of the recession, the company has managed to prove that it can "move forward and grow in difficult times" increasing the turnover, creating jobs and boosting company's productivity (Mercadona, 2010). The strength of Total Quality Management Model and the tight links that create between firm's stakeholders has allowed the firm to effectively and efficiently respond to the lost of clients caused by the economic crisis in 2008 transforming the crisis from a threat to firm survival to an opportunity to increase its market share and consolidate its competitive advantage over competitors.

6. Acknowledgement

This chapter has been supported by Project ECO2009-10358 of Spanish Ministry of Science and Innovation (Spain) and Catedra Iberdrola for Research in Business Management and Organization.

7. References

Alonso, M.; Furio, E. (2010). La economía española del crecimiento a la crisis pasando por la burbuja inmobiliaria, *Cahiers de civilisation espagnole contemporaine*, Vol 6, Available from http://ccec.revues.org/index3212.html

Blanco, M.; Gutierrez, S. (2010). Application of the Total Quality Management Approach in a Spanish Retailer: The Case of Mercadona. *Total Quality Management and Business Excellence*, Vol.21, No.12, (December 2010), pp.1365 – 1381

Caparrós, A. & Biot , R. (2006). *15 x 15, quince días con quince empresarios líderes*. Fundación para la Comunidad Valenciana - Escuela de Dirección de Empresas del Mediterráneo (EDEM), Valencia, España

Deloitte (2007). *Global Powers of Retailing 2006*, New York, USA

ÉTNOR (2003). *Informe de Auditoría Ética*. Fundación para la ética de los negocios y las administraciones, Madrid, España

Hanna, J. (2010). How Mercadona Fixes Retail's "Last 10 Yards" Problem, *Research&Ideas, Harvard Business School Working Knowledge*, Cambridge, MA.

INE (2011). Instituto Nacional de Estadística (National Statistics Institute). Data available from http://www.ine.es/

Lagrosen, S. (2001). Strengthening the weakest link of TQM – from customer focus to customer understanding. *The TQM Magazine*, Vol.13, No. 5, pp. 348-354.

Mercadona. (1999-2010). Mercadona's Annual Reports. Tavernes Blanques, Valencia, España

Navarro, R. (2005). *Los nuevos burgueses valencianos*., La esfera de los libros, Valencia, España

Reputation Institute (2007). *Global Rep. Track Pulse 2007*. Reputation Institute, New York, USA

Ton, Z.; Harrow, S. (2010). Mercadona. *Harvard Business School Cases*. 9-610-089. Cambridge, MA, USA

Wicks, A.C. (2001). *The Value Dynamics of Total Quality Management: Ethics and the Foundations of TQM*. Business Ethics Quarterly, Vol.11, No.3, pp. 501-536.

13

The Relationship Between ISO 17025 Quality Management System Accreditation and Laboratory Performance

Esin Sadikoglu and Talha Temur
Gebze Institute of Technology,
Turkish Standard Institution,
Turkey

1. Introduction

International Organization for Standardization ISO 9000 is a general standard related to fundamentals of implementation of a quality management system. The aim of ISO 9000 is to implement quality management procedures by means of leadership, work instructions, documentation and record keeping. ISO 9000 standard does not determine the quality or performance level of the product or service. It deals with standardization of process management and improvement of an organization. The origin of ISO 9000 standards comes from harmonizing quality standards across nations (Heizer & Render, 2011; Goetsch & Davis, 2010).

Laboratory accreditation assesses the competencies of all types of laboratories in terms of performing specific tests and calibrations. ISO and the international Electro-Technical Commission (IEC) introduced ISO/IEC 17025 standard due to the growing importance of accreditation and international recognition. The assessment of the facilities that calibrate and test equipment is crucial to monitor accuracy of measurement and testing (Beckett and Slay, 2011). ISO 17025 is a laboratory standard equivalent to more generic ISO 9000. The standard can be applied by all organizations performing tests and/or calibrations and it can harmonize laboratories worldwide (ISO-9000, 2010).

ISO 17025, as all ISO standards, heavily focuses on documenting the process of any analysis performed by a laboratory. It includes the quality management system and technical requirements of the accreditation process. The management requirements section of the standard evaluates the organization of the laboratory; its quality system; document control; review of requests; tenders and contracts; subcontracting of tests and calibrations; purchasing services and supplies; service to the customer; complaints; control of nonconforming testing and/or calibration work; corrective actions; preventive actions; improvement; control of records, internal audits and management reviews. This section adapts the ISO 9000 quality management criteria into a laboratory context. The technical requirements section of the ISO 17025 standard evaluates personnel; accommodation and environmental conditions; test and calibration methods and method validation; equipment; measuring traceability; sampling; handling of test and calibration items; assurance of the quality of test and calibration results, and reporting the results. A laboratory must identify both the management requirements and technical requirements of the standard in order to produce a good product and satisfy its customer (ISO/IEC-17025, 2010). Although ISO standards and total quality management implementations have the same objective of improving competitiveness of the organization, they are not interchangeable. They are compatible with and support each other. However, the majority of ISO certified organizations do not fully implement total quality (Goetsch & Davis, 2010).

As with ISO 9000 (Goetsch & Davis, 2010), ISO 17025 is not prescriptive in its content. In other words, the standard does not insist a laboratory pursue the methods nor does it determine the qualifications of the staff to perform a test or calibration. It lets the laboratory determine and document its own standards needed. Then the standard ensures the laboratory comply with the standards set by the laboratory (Beckett and Slay, 2011).

There is a need to compare the laboratories' experimental measurement results for monitoring and assessing their performance. However, there are limited studies as well as ambiguous and mixed results of the relationship between ISO 17025 accreditation and laboratory performance in literature. This study contributed to figure out the effect of the adoption of ISO 17025 accreditation of a laboratory on its performance. The paper is organized as follows: Section 1 introduces ISO 17025 standard and the importance of study. Section 2 presents previous studies about ISO 17025 accreditation implementations and laboratory performance. Section 3 gives methodology of the study including data collection and methods of analysis. Section 4 presents results of the analysis, discussion, and implications of the study. Final section concludes the paper.

2. Literature review

Laboratory accreditation can help laboratories to produce consistent results by means of implementing the framework of a documented quality system (Beckett and Slay, 2011).

Cooper, Moeller, & Kronstrand (2008) presented the current status of accreditation of hair testing laboratories based in Europe. They found 48 percent of the laboratories accredited and 31 percent of the laboratories accredited to ISO 17025.

Malkoc & Neuteboom (2007) found approximately 67 percent of government or police laboratories were not accredited. They claimed there was ambiguity of perceptions about the meaning, purpose, and principles of quality assurance and accreditation among European forensic science laboratories. The personnel of some laboratories did not know the fact that a laboratory without having a quality management system cannot get accredited. This problem is attributed to "lack of awareness syndrome."

There are mixed results about the relationship between ISO 17025 accreditation and laboratory performance. Zaretzky (2008) claims the improper use of quality management tools in the laboratory has a negative effect on laboratory activity especially on continous improvement implementation. This also always a negative effect on the customer.

Guntinas et al. (2009) performed proficiency test for heavy metals in feed and food of 31 laboratories, 28 of which were ISO 17025 accredited laboratories, in Europe. They found low laboratory performance and claimed the errors in measurement uncertainity might be the reason of failure of the laboratories.

Raposo et al. (2009) conducted proficiency testing of chemical oxygen demand measurements and found low laboratory performance. They claimed the failure might come from heterojenity of the samples, meaurement errors, and the differences in the analysis methods.

Korun & Glavic-Cindro (2009) made an analysis of the causes of discrepant results in proficiency tests in a testing laboratory and investigated the reasons of erroneous laboratory results in the years of 2003 and 2007.

Ahmad, Khan, & Ahmad (2009) investigated accreditation process and standardization of Pathology Laboratories in Pakistan. They claimed physical conditions and limited qualified workforce were the reasons of current insufficient accreditation status in Pakistan.

Uras (2009) exlained quality regulations and accreditation standards for clinical chemistry in Turkey. He mentioned insufficient accreditation status in clinic laboratories and emphasized the importance of having a necessary infrastructure and directions in the healhtcare in Turkey.

On the other hand, there are successful implementations of accreditation reported in the literature. Iglicki, Mila, Furnari, Arenillas, Cerutti, & Carballido (2006) showed accreditation process of the Quality System by ISO 17025 standards in a national reference laboratory of Argentina. They found the implementation improved systematic recording and control of the tasks, robustness of the traceability chain and external recognition of quality of the laboratory.

Cortez (1999) found accredited laboratories had more satisfactory and less suspicious and unsatisfactory laboratory performances than non-accredited laboratories. Thus, they claimed accredited laboratories were more succesful than non-accredited laboratories.

Hall, Maynard, & Foster (2003) have explained implementations of ISO 17025 accreditation of a laboratory within racing chemistry in Britain. They reported successful results in spite of difficulties of implementation of accreditation.

Vlachos, Michail, & Sotiropoulou (2002) presented implementation and maintance of the ISO 17025 quality assurance system in the General Chemical State Laboratory of Greece. The laboratory could prove the reliability of the test results and technical competence to clients and regulators. They experienced that accreditation and international quality standard of the laboratory improved its competitiveness. The laboratory could fulfill the requirements of clients and regulators. The laboratory could effectively analyze, solve and reduce operational problems with the adoption of a quality culture. The laboratory obtained employee fulfillment with the personnel training and teamwork. Based on the reviewed literature, we propose the following hypothesis:

H1: The existence of ISO 17025 accreditation affects laboratory performance.

We tested this hypothesis in all laboratories, as well as in each sector namely manufacturing, private, state, building materials, chemical, electro-magnetic conformity (EMC), metallic material and textile laboratories, separately.

There were 1 million ISO 9000 certified firms in 175 countries by 2009 (Heizer & Render, 2011). Table 1 and Figure 1 shows accreditation status of countries with respect to the number of accredited laboratories per million people of each country as of 2010. The numbers of accredited laboratories were obtained from the web pages of the national accreditation institutions of the countries given in the table and figure. As clearly noticed in the table and figure, Australia (98.3) and Singapore (84.3) have the most accredited laboratories and Turkey (4.4) has the least accredited laboratories per million people.

Country	Number of accredited laboratories	Population (million)	Laboratory/population (#/million people)
Turkey	316	72.0	4.4
Australia	2065	21.0	98,3
Denmark	160	5.5	29.1
France	1700	64.0	26.6
Germany	2290	82.0	27.9
Hong Kong	170	7.0	24.3
Norway	145	4.5	32,2
Singapore	295	3.5	84,3
Spain	780	41.0	19,0
Czech Republic	610	10.0	61,0
England	1970	61.0	32,3
Italy	926	58.0	16,0
Sweden	585	9.0	65,0

Table 1. Accreditation status of some countries.

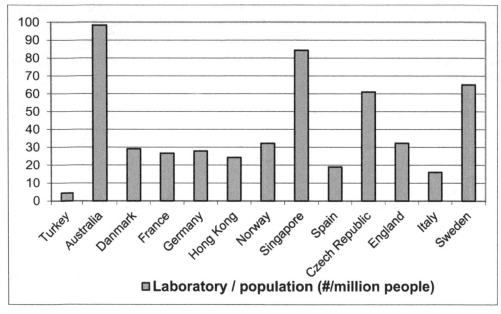

Fig. 1. Accreditation status of some countries.

3. Methodology

Experiment Laboratory Center of the Turkish Standards Institution is the only member of the European Proficiency Testing Information system (EPTIS) (TSE, 2010). The study was informed in the web pages of Turkish Standards Institution as well as by telephone and e-mail to laboratories in March 2009. The study used 36 experiment program, some of which require multiple laboratory measurements. For example, Experiment Program 8 required eight different laboratory measures.

Laboratories were classified with respect to organizational structure namely, manufacturing, private, and state laboratories for the analysis. They were also classified with respect to industry sector, namely building materials, chemical, electro-magnetic conformity (EMC), metallic materials and textile laboratories.

Laboratory performance measures were standardized for the analysis since the measures were different in different industry sectors. The laboratory performance measures were classified as satisfactory ($|z| \leq 2$), suspicious ($2 < |z| < 3$) and unsatisfactory ($|z| \geq 3$) to compare laboratories' frequencies of performances. The absolute standard values of the laboratory measurements of greater than 10 ($|z| > 10$) were discarded for improving reliability. Grubbs test was used for determining and discarding the extreme values from the analysis. In addition, Cochran test was used for confirming homogeneity of variance of laboratories before conducting t-test (Ferrero & Casaril, 2009).

Data were analyzed with frequency comparison; t-test and Welch test with respect to existence of ISO 17025 accreditation of the laboratories. The assumptions of the independent-sample t-test were satisfied before the analysis. For this purpose, Kolmogorov-Smirnov test was used for testing normality of the data and Levene's test was used for testing homogeneity of variance of the data. Welch test was conducted to compare means of the two groups when homogeneity of variance assumption was not satisfied according to Levene's test results.

4. Results of the analysis and discussion

The study obtained 564 laboratory measurements from participating 73 laboratories in different sectors located 20 cities in Turkey. Most of the participant laboratories (68.5%) conducted one to three experiment programs. The participant laboratories represented Turkish economy since the regional dispersion of the participants was similar to the entrepreneur (manufacturing) numbers according to business registration in 2008 (TUIK, 2010).

Most of the participant laboratories were manufacturing laboratories (46.6%) and state laboratories (32.9%). The number of ISO accredited laboratories (31, 42.5%) was close to the number of non-ISO accredited laboratories (34, 46.6%). Most of the ISO accredited laboratories (58.1%) were state laboratories while most of non-ISO accredited laboratories (73.5%) were manufacturing laboratories.

Kolmogorov-Smirnov test results confirm that normality assumptions were satisfied. Table 2 gives comparison results of laboratory performance percentages with respect to ISO 17025 accreditation existence. Data in the table are in percentages. The first data in the cells are satisfactory laboratory performance ($|z| \leq 2$) percentages and the second data given in the parentheses are unsatisfactory laboratory performance ($|z| \geq 3$) percentages. Table 3 gives independent sample t-test or Welch test results of laboratory performance.

Table 2 shows that satisfactory laboratory performances of ISO accredited laboratories range from 71.7 to 100 percent and their unsatisfactory laboratory performances range from 0 to 20.8 percent. On the other hand, satisfactory laboratory performances of non-ISO accredited laboratories range from 65.2 to 100 percent. Their unsatisfactory laboratory performances range from 0 to 11.1 percent.

Laboratories	ISO 17025 Accreditation Existence	
	ISO Accredited (%)	Non-ISO Accredited (%)
All laboratories	$85.8^a(11.7^b)$	86.3 (7.7)
Manufacturing laboratories	83.1 (16.9)	91.4 (6.7)
Private laboratories	71.7 (20.8)	65.2 (4.3)
State laboratories	94.8 (2.9)	86.1 (11.1)
Building materials laboratories	92.5 (5.0)	88.1 (9.5)
Chemical laboratories	100.0 (0.0)	87.5 (8.3)
EMC laboratories	78.1 (19.5)	78.0 (4.9)
Metallic materials laboratories	88.0 (8.0)	91.7 (8.3)
Textile laboratories	97.2 (0.0)	100.0 (0.0)

a. Satisfactory laboratory performance ($|z| \le 2$) percentage
b. Unsatisfactory laboratory performance ($|z| \ge 3$) percentage

Table 2. Laboratory performance percentages comparison results.

As clearly noticed in Tables 2 and 3, the existence of ISO 17025 accreditation of a laboratory statistically significantly affects laboratory performance in all laboratories, manufacturing laboratories and EMC laboratories in a negative way. In other words, non-ISO 17025 accredited laboratories, in general, have higher satisfactory laboratory performance ($|z| \le 2$) and lower unsatisfactory laboratory performance ($|z| \ge 3$) than ISO 17025 accredited laboratories in all, manufacturing and EMC laboratories. Furthermore, existence of ISO 17025 accreditation does not statistically significantly affect laboratory performance in the other sectors.

Laboratories	Statistics	Degree of freedom	Significance (two-tailed) level	State of the hypothesis
All	2.86^a	358.07	0.09	**Supported**
Manufacturing	2.97^a	64.05	0.09	**Supported**
Private	0.75^b	74.00	0.46	Not supported
State	-0.67^b	208.00	0.50	Not supported
Building materials	0.97^b	122.00	0.33	Not supported
Chemical	0.68^a	33.71	0.42	Not supported
EMC	3.74^a	114.97	0.06	**Supported**
Metallic materials	-1.06^b	35.00	0.30	Not supported
Textile	-1.49^b	76.00	0.14	Not supported

a. Asymptotically F distributed; Welch test was used.
b. Independent samples t-test was used.

Table 3. Independent sample t-test or Welch test results of laboratory performance.

The results (Tables 2 and 3) show that all ISO accredited laboratories, in general, have lower satisfactory laboratory measures (85.8%) and higher unsatisfactory laboratory measures (11.7%) than non-ISO accredited laboratories (86.3% and 7.7%, respectively).

Furthermore, ISO 17025 accredited manufacturing laboratories have, in general, produce less satisfactory laboratory measures (83.1%) and more unsatisfactory laboratory measures (16.9%) than non-ISO 17025 accredited manufacturing laboratories (91.4% and 6.7%, respectively). Results imply that manufacturing laboratories cause the negative effect of ISO 17025 accreditation on laboratory performance. Manufacturing laboratories measure the products of their host companies. Thus, ISO accredited manufacturing laboratories may not be careful in measurement and they may give more erroneous results than non-ISO accredited manufacturing laboratories.

Moreover, ISO 17025 accredited EMC laboratories, in general, give more unsatisfactory laboratory measures (19.5%) than non-ISO accredited EMC laboratories (4.9%) although their performances of satisfactory laboratory measures are close to each other.

ISO certification has some benefits: Periodic audits give pressure the firms to maintain conformance to the requirements of the standard. In addition, certified organizations will get more recognition and credibility than non-ISO certified organizations in the world's marketplace. Documentation can confirm consistency in an organization (Maynard et al., 2003). On the other hand, ensuring consistency in quality at any level via ISO 9000 or ISO 17025 does not mean high quality. The procedures of the ISO standard do not show actual quality of the product or service (Goetsch & Davis, 2010; Vlachos, Michail, & Sotiropoulou, 2002; Heizer & Render, 2011). Thus, the ISO accreditation can be seen as a thereshold or a minimum quality level to be accepted in the market.

The negative or insignificant effect of ISO 17025 accreditation existence on laboratory performance may come from the reason of acquiring the accreditation being customer or supplier requirement, advertisement, and marketing gimmick purposes. ISO 17025 accredited laboratories may not give importance to customer satisfaction, employee fulfillment and quality of the product, service or process.

Laboratories who attempt to accreditation under the ISO 17025 standard face challenges of education and validation of tools and methods (Beckett and Slay, 2011). Top management support is important to improve employee fulfillment, consistency, reliability and accuracy of laboratory measurements, quality performance and customer satisfaction. Providing necessary training to employees can increase awareness about and commitment to produce quality products. Therefore, adequate training on quality and management support should be given to manufacturing and EMC laboratories to improve quality and laboratory performance.

The results of the analysis were discussed with some representatives of the participant laboratories and designers of the experiments. They claim that most laboratories try to implement ISO 17025 accreditation of quality management system. On the other hand, the laboratories with low endorsement avoid ISO 17025 accreditation due to its high financial cost. This result is consistent with Cooper, Moeller, & Kronstrand's (2008) claim. Closely monitoring design, planning and implementation stages of ISO 17025 accreditation of quality management system for three to four years is highly recommended for practitioners.

Globalization increases the need for credibility of the laboratory. ISO 17025 accreditation can improve recognition. On the other hand, achieving accreditation requires considerable amount of financial resources for laboratories. Furthermore, the quality assurance system implementation can increase bureaucracy through complicate updated records. It requires extensive human resources tasks. Thus, it is difficult, time-consuming and costly (Vlachos, Michail, & Sotiropoulou, 2002).

The extensive huge quantity of documentation developed for fulfilling the requirement of the standard can cause problem of redundancies of information in multiple documents. This leads to contradictory directives to its users. Especially, process changes should be modified in related multiple documents but the corrections are not usually done in every document. This causes inconsistency of the information, contradiction, and thus, important non-conformities in the system. If the members of the personnel follow different instruction for the same task, the performance of the laboratories will reduce (Zaretzky, 2008). When a process is performed incorrectly, technical validity of the results of the laboratory will reduce (Maynard, Foster, & Hall, 2003). These problems can be remedied by implementing an enterprice resource planning (ERP) system with a common database. Monitoring documents, updating changes in the process, maintaining reliable, consistent, timely and accurate data and information throughout the laboratory can be achieved efficiently and effectively with an ERP system as well.

Employees of the laboratories can resist to the quality management implementation since their laboratories have already performed their technical work in a proper way before implementing the ISO 17025 Standard. Thus, employees may not commit to the requirements of the standard (Zaretzky, 2008).

The important factors for laboratories in seeking accreditation are motivation and determination (c.f. Malkoc & Neuteboom, 2007). Top management commitment to quality is crucial to give adequate resources and to motivate the employees for their participation. Training should improve knowledge and skills of the employees for imlementing of quality management and accreditation successfully. There should be a common understanding of accreditation, its purpose and its process in order to get the maximum benefit from the efforts of accreditation.

Certification and accreditation may provide recognition and reputation of the laboratory both in their country and in the global economy. However, if a laboratory applies to ISO 17025 certification for solely marketing purpose without total quality management focus, it will eventually lose their clients, credibility, competitiveness, and profitability.

The main reason for adopting accreditation is the marketing strategy due to high financial responsibility of being accredited (Cooper et al., 2008; Goetsch & Davis, 2010). Implementing quality management in laboratories brings additional time, effort and cost as this happens to all organizations. Some laboratories may be ISO 17025 accredited for merely gaining a marketing advantage rather than other motivations such as improving reliability, accuracy and consistency of their laboratory measurements. Such accredited laboratories do not get these benefits of ISO adoption due to their inappropriate motivations for adopting the accreditation.

The results of the study are supported by (Zaretzky, 2008)'s and (Uras, 2009)'s claims. The reasons of failure of ISO 17025 accreditation implementations of the laboratories in Turkey for improving quality, effectiveness, and performance of laboratories may be as follows: There may be errors in the auditing by the registrar of the laboratories in Turkey. Some certification bodies and their personnel may overlook the evaluation process instead of auditing with respect to the requirements of ISO 17025 accreditation in detail. In addition, some certification bodies and their personnel may not be honest and impartial in auditing. These can result in issuing the accreditation to the laboratories unfairly. Thus, certification bodies and their personnel may not audit effectively or ethically. Some certification bodies and the laboratories in Turkey may lack ethical guidelines in the audits and the accreditation process implementation to pursue their missions. In addition, some laboratories may adopt ISO 17025 accreditation for solely gaining a marketing advantage without total quality management focus. Thus, some laboratories may get ISO 17025 accreditation although they are not qualified with respect to the requirements of the standard. Therefore, we suggest that conformity assessments of the registrars (i.e. certification bodies' organizations, personnel, and procedures) in auditing and awarding the certificate should be conducted more strictly and frequently by the accreditation body to minimize errors and to improve impartiality, consistency, effectiveness and accuracy of the awarding ISO 17025 accreditation. This will also ensure and improve credibility of the ISO 17025 accreditation. Moreover, we propose that all stakeholders in the audit and accreditation process implementation should give more importance to ethics, quality and continual improvement in their guiding principles so that the laboratories will improve their performances by means of adopting the ISO accreditation.

When the effect ISO 17025 accreditation on laboratory performance is examined in other countries, fruitful discussions and suggestions can be generated. Thus, we recommend researchers to reexamine the effect of ISO 17025 accreditation on performance in other countries. In addition, longitudinal study may provide trends in the relationship between ISO 17025 accreditation and laboratory performance.

5. Conclusion

This study examined the effect of ISO 17025 accreditation on laboratory performance in Turkey. The results show statistically negative or insignificant effects on laboratory performance. When a laboratory adopts ISO 17025 accreditation for merely gaining a marketing advantage, this wrong motivation brings the laboratory negative outcomes such as high unsatisfactory and suspicious laboratory measurements and low satisfactory laboratory measurements. On the other hand, the laboratory adopts ISO accreditation for such purposes of improving quality, reliability, accuracy and consistency of its products, services and processes, and customer satisfaction, the laboratory can improve its laboratory performance, customers' loyalty and competitiveness in the market.

We suggest that the accreditation body should control and assess competency and honesty of the organization and personnel of the certification bodies in audits more strictly and frequently to improve credibility, fairness, and effectiveness of ISO accreditation. In addition, we suggest that the accreditation body, certification bodies, and laboratories should more emphasize such issues as ethics, quality, and continual improvement in their guiding principles to improve the impartiality and competency of the audits as well as to have good motivations for ISO adoption, and to increase quality, reliability, accuracy, and consistency of laboratory measurements.

6. References

Ahmad, M., Khan, F. A., & Ahmad, S. A. (2009). Standardization of Pathology Laboratories in Pakistan: Problems and Prospects. *Clinical Biochemistry* , 42, 259-262.

Cooper, G., Moeller, M., & Kronstrand, R. (2008). Current Status of Accreditation for Drug Testing in Hair. *Forensic Science International* , *176*, 9-12.

Cortez, L. (1999). The implementation of accreditation in a Chemical Laboratory. *Trends in Analytical Chemistry* , *18* (9-10).

Ferrero, C., & Casaril, M. (2009). Proficiency Testing Programs to Improve Traceability in Chemical Analysis. *Measurement* , *42*, s. 1502-1509.

Goetsch, D. L., & Davis, S. B. (2010). *Quality management for organizational excellence-introduction to total quality* (sixth ed. b.). New Jersey: Prentice Hall.

Guntinas, M. B., De La, C., Wysocka, I., Quetel, C., Vassileva, E., Robouch, P., et al. (2009). Proficiency Test for Heavy Metals in Feed and Food in Europe. *Trends in analytical Chemistry* , *28* (4).

Hall, D. J., Maynard, S., & Foster, S. (2003). ISO 17025 Application within Racing Chemistry: A Case Study. *Technovation* , *23*, 773-780.

Heizer, J., & Render, B. (2011). *Operations Management.* New Jersey: Prentice Hall.

Iglicki, A., Mila, M. I., Furnari, J. C., Arenillas, P., Cerutti, G., & Carballido, M. (2006). Accreditation Experience of Radioisotope Metroloji Laboratory of Argentina. *applied Radiation and Isotopes* , *64*, 1171-1173.

ILAC *(International Laboratory Accreditation Cooperation).* (2010). htpp://www.ilac.org

ISO/IEC-17025. (2010). General Requirements of Testing and Calibration Laboratories. Geneva, Switzerland: International Organization for standardization.

ISO-13528. (2005). Statistical Methods for Use in Proficiency Testing by Interlibrary Comparisons.

ISO-9000 . (2010). Quality Management Systems: Fundamentals and Vocabulary: htpp://www.iso.org

Korun, M., & Glavic-Cindro, D. (2009). An Analysis of the Causes of Discrepant Results in Proficiency Tests in a Testing Laboratory. *Applied Radiation and Isotopes* , *67*, 683-686.

Malkoc, E., & Neuteboom, W. (2007). The Current Status of Forensic Science Laboratory Accreditation in Europe. *Forensic Science International* , *167*, 121-126.

Maynard, S., Foster, S., & Hall, D. J. (2003). ISO 17025 Application within Racing Chemistry: A Case Study. *Technovation* , *23*, 773-780.

Raposo, F., Rubia, M. A., Borja, R., Alaiz, M., Beltran, J., Cavinato, C., et al. (2009). An Interlaboratory Study as Useful Tool for Proficiency Testing of Chemical Oxygen Demand Measurements Using Solid Substrates and Liquid Samples with High Suspended Solid Content. *Talanta* , *80*, 329-337.

TSE. (2010). *Turkish Standard Institution.* htpp://www.tse.org.tr

TSE. (2010). *Turkish Standard Institution Inter-laboratory Comparison Test Program- 2009 Final Report.* Gebze.

TUIK. (2010). *Turkish Statistics Association.* htcp://www.tuik.gov.tr

TURKAK. (2010). *Turkish Standard Institution Inter-laboratory Comparison Test Program.* htpp://www.turkak.org.tr

Uras, F. (2009). Quality Regulations and Accreditation Standards for Clinical Chemistry in Turkey. *Clinical Biochemistry* , *42*, 263-265.

Vlachos, N. A., Michail, C., & Sotiropoulou, D. (2002). Is ISO/IEC 17025 Accreditation a Benefit or Hindrance to Testing Laboratories? The Greek Experience. *Journal of Food Composition and Analysis* , *15*, 749-757.

Zaretzky, A. N. (2008). Quality management systems from the perspective of organization of complex systems. *Mathematical and Computer Modelling* , *48*, 1170-1177.

14

The Role of Person-Organization Fit in TQM: Influence of Values and Value Congruence on TQM Orientation

Alper Ertürk
Vlerick Leuven Gent Management School
Belgium

1. Introduction

Despite the fact that total quality management (TQM) philosophy, approach and practices have widely been adopted by a substantial number of organizations in the last decades, the evidence of its positive influence on organizational-level and individual-level performance variables is somewhat mixed (Choi & Behling, 1997; Coyle-Shapiro & Morrow, 2003). When it comes to explaining why TQM initiatives succeed or fail, usually the elements, such as process management, implementation issues, training, top management support, as well as technical and cultural issues, have been cited (e.g. Reger et al., 1994). However, since TQM has been defined as a system level management approach (Sitkin, Sutcliffe & Schroeder, 1994), varying attitudes and behaviors of individual employees, as well as the specific factors that may differentially effect individual variability, have seldom been considered and have not been emphasized in TQM literature (Coyle-Shapiro & Morrow, 2003).

On the other hand, since individuals cannot be isolated from their work environment, researchers have started to emphasize the concept of person-environment fit, and its significant influence on various work-related outcomes (Kristof-Brown, Zimmerman & Johnson, 2005). In its most general sense, person-environment (P-E) fit can be defined as the congruence, match, similarity, or correspondence between the person and the environment. Within this general definition, several different dimensions have evolved to conceptualize person environment fit (Kristof, 1996). The first and the most important dimension, which is focused in this study, is the supplementary versus complementary distinction. Supplementary fit occurs when a person possesses characteristics, which are similar to others in an environment, whereas complementary fit occurs when a person possesses different characteristics than others, which add to the environment what is missing.

Person-organization fit (P-O Fit) is an important and broadly studied type of P-E fit. In general terms, P-O fit concerns the antecedents and consequences of the compatibility between individuals and the organization in which they work (Kristof, 1996). The crucial importance of a good fit between the individual and the organization has comprehensively been stressed by several researchers and practitioners, who have contended that P-O fit is a key factor in maintaining a committed and flexible workforce that is necessary in a competitive business environment (Kristof, 1996; Coyle-Shapiro & Morrow, 2003). Empirical evidence yielded that high level of P-O fit is related to positive attitudinal and behavioral outcomes (Bretz & Judge, 1994; O'Reilly, Chatman & Caldwell, 1991; Posner, 1992).

Although TQM is an organization-wide system level approach and management philosophy, perception and adoption of quality orientation, as well as involvement and participation to the quality oriented practices, are individual level constructs. Thus, having congruence between the individuals and their organizations can be considered as a very important factor to enhance the adoption of quality orientation. However, to the best knowledge of the author, no research to date has focused on the role and influence of P-O fit as value congruence on the employees' adoption of TQM orientation. Therefore, this study is undertaken to bridge the gap in the literature by focusing on the influence of supplementary fit between employees work values and organizational values on the adoption of TQM orientation. The current research addresses linear associations of personal and organizational work values, as well as curvilinear relationship of the significant interactions of those value dimensions on TQM orientation. Main purpose of this research is to determine how P-O fit, in terms of personal and organizational values and their congruence and discrepancies, influences the adoption of TQM orientation, which includes participation in teamwork and involvement in continuous improvement. To examine those linear and curvilinear associations among various components, polynomial regression and surface response analyses were employed. Furthermore, three-dimensional graphs were also provided to better explain the proposed and yielded results.

This study contributes to literature in several aspects. First, unlike most research, which examines TQM orientation as an independent variable, this study specifies this concept as a dependent variable. Such an approach elevates current work as it responds to the need of research examining factors that influence adoption of quality orientation and consequently application of TQM approach. Second, instead of only investigating the linear effects of personal and organizational values on adoption of quality orientation separately, this study focuses on a rather neglected aspect of the P-O fit by examining the combined curvilinear effects of those values. In particular, this study posits that personal and organizational values have significant relationship with TQM orientation, as well as congruence and discrepancies between those values influence TQM orientation in a non-linear way. Despite the widely acknowledged importance of P-O fit theory, prior attempts have not generated sufficient understanding of its importance in implementation of TQM approach. Third, the results of this study are based on polynomial regression and surface response analyses in order to better examine the significant linear and curvilinear relationships among constructs, and to overcome methodological problems related to difference scores and other traditional congruence measures (e.g., profile correlations), which are commonly used in value congruence research. Finally and most importantly, despite the importance of individual differentiations in TQM implementations, to our best knowledge, this study is among the first attempts to explore P-O fit components as antecedents of adoption of TQM orientation.

2. Literature review and theoretical framework

2.1 Total Quality Management (TQM) orientation

The definition of TQM did not evolve as a result of the academic analysis of existing management literature and organizational theory. It has been mainly built on the practice-oriented studies of Juran (1989), Deming (1986), and Ishikawa (1985). Based on the initial success of applying statistical process control and systematic planning, these researchers included in their definitions issues such as employee motivation, teamwork, continuous improvement, as well as the relations between customer satisfaction and economic results. Hence, TQM has a solid foundation of statistical thinking, along with a number of prescriptive management ideas. Lately, TQM programs go beyond implementing technical solutions and management practices, which focus on organizational members working together to meet customer requirements (Hackman & Wageman, 1995). Thus, TQM cannot be analyzed by just focusing on visible technical interventions, but by gaining a comprehensive understanding of the underlying assumptions of individuals, which are the key components for the success of those interventions.

TQM orientation, on the other hand, can generally be defined as the employees' awareness of the importance of quality and felt moral responsibility for achieving it (Coyle-Shapiro & Morrow, 2003). It broadly shows to what extent employees support and adapt to the TQM initiatives. Most research in TQM literature identify mainly three core components for TQM orientation, which are customer satisfaction, continuous improvement and teamwork (e.g., Dean & Bowen, 1994; Coyle-Shapiro & Morrow, 2003). However in this study, since we specifically focus on employees' individual differences and congruence with the organization, we include teamwork and continuous improvement, which together constitute TQM orientation. Teamwork has broadly been defined as employees' willingness to collaborate and cooperate in different levels from work group to interorganizational activities, as emphasizing the quality of group functioning within TQM framework (Dean & Bowen, 1994). Continuous improvement has been defined as the recognition of and felt responsibility for quality improvement and involvement in quality enhancing activities and TQM practices (Coyle-Shapiro & Morrow, 2003). Continuous improvement suggests that employees continuously change their thinking and the way of doing business, which they adapt in their work.

Even though TQM is considered as an organization-wide approach, TQM orientation as involvement and participation to the quality oriented practices is an individual level construct. Nevertheless, surprisingly little research on TQM has emphasized individual differences and variations (Coyle-Shapiro & Morrow, 2003). Coyle-Shapiro and Morrow (2003) also stated that the achievement of TQM goals would be limited without considering individual-level factors, such as values and attitudes. An important model that takes into consideration individual differences within the framework of TQM orientation is person-organization fit. According to the P-O fit model, desirable outcomes can be achieved when employees' values, abilities or goals are congruent with the organizational characteristics. Hence, examining the extent to which employees' individual differences or congruences with the organization would substantially contribute to our current understanding and generalizability of TQM and P-O fit models.

2.2 Person-Organization (P-O) fit

Person-organization (P-O) fit has received a considerable attention from researchers and practitioners as being one of the most important workplace variables. P-O fit has been defined as the extent to which an employee's characteristics are compatible with organizational characteristics (Kristof, 1996). One of the fundamental definitions of P-O fit has been proposed by Kristof (1996), as "the compatibility of people and organizations that occurs when (a) at least one entity provides what the other needs, or (b) they share similar fundamental characteristics, or (c) both" (p. 271).

Empirical evidence suggests that P-O fit predicts many positive work-related attitudes and behaviors (Kristof, 1996; Verquer, Beehr & Wagner, 2003). For example, research has demonstrated that higher levels of P-O fit most likely results in higher levels of job satisfaction (e.g., Kristof-Brown et al., 2005; Chatman, 1989), organizational commitment (e.g., Cable & Judge, 1996; Bretz & Judge, 1994), organizational citizenship behaviors (O'Reilly et al., 1991; Ucanok, 2009), teamwork and performance (Posner, 1992), and organizational identification (e.g., Amos & Weathington, 2008), as well as lower level of turnover intentions (e.g., Saks & Ashforth, 2002; Cable & De Rue, 2002). In addition, P-O fit has been also found to influence job search and job choice decisions (Cable & Judge, 1996), along with recruiter perceptions of applicant suitability (Bretz & Judge, 1994).

In the literature, P-O fit has been conceptualized in different ways. Four commonly accepted ways of defining P-O fit include supplementary fit (i.e., when an individual possesses characteristics that are similar to existing organizational characteristics), complementary fit (i.e., when an individual fills a void or adds something missing in the organization), needs-supplies fit (i.e., when an individual's needs are fulfilled by an organization), and demands-abilities fit (i.e., when an individual's abilities meet the demands of the organization) (Muchinsky & Monahan, 1987; De Clerq et al., 2008).

Supplementary fit occurs when an employee possesses characteristics that are similar to characteristics existing in the organizational environment (Muchinsky & Monahan, 1987). In other words, when an employee's characteristics are congruent to organizational characteristics, a supplementary fit exists. On the other hand, complementary fit occurs when an employee's characteristics complement the characteristics of organizational environment (Muchinsky & Monahan, 1987). Thus, an employee has a kind of complementary fit when he or she possesses different characteristics that add to existing organizational characteristics. Needs-supplies fit occurs when an employee's needs are fulfilled by the organization (Caplan, 1987). Hence, an employee has needs-supplies fit to the extent that the organization satisfies of this individual's needs or preferences. Nevertheless, demands-abilities fit reflects the degree to which an employee's abilities fit the demands of the organization (Caplan, 1987). According to this model, demands-abilities fit occurs when an employee possesses characteristics that are required for meeting organizational demands. In this study, we use supplementary fit, since supplementary fit has been empirically proven to be one of the most widely used P-O fit concepts as being associated with various work-related outcomes (Kristof, 1996; Ostroff, Shin & Kinicki, 2005).

2.3 Values and value congruence as P-O Fit

A substantial amount of research has emphasized the importance of congruence between the values of employees and organizations (Edwards & Cable, 2009; Amos & Weathington, 2008; Chatman, 1989; Kristof, 1996; Ostroff et al., 2005). Consistent with prior research (De Clerq, Fontaine & Anseel, 2008; Schwartz, 1992), in this study we define values as general beliefs about the importance of normatively desirable behaviors. Values hold by employees guide employees' actions and decisions, whereas organizational value systems provide norms that define how organizational members are expected to behave and how organizational resources should be allocated.

Schwartz (1992) has proposed a comprehensive theory about the content of individual value systems. His systematic value theory has also been empirically validated (e.g., Schwartz et al., 2001). Individual value dimensions of Schwartz model are presented in Table 1.

Value	Definition
Conformity	Restraint of actions, inclinations, and impulses likely to upset or harm others and violate social expectations or norms
Goal-orientedness	Living and working to fulfill a purpose, not giving up
Hedonism	Pleasure and sensuous gratifications for oneself
Materialism	Attaching important to material goods, wealth and luxury
Nature	Appreciation, preservation and protection of nature
Power	Control or dominance over people
Prestige	Striving for admiration and recognition
Relations	Having good interpersonal relations with other people and valuing true friendship
Security	Safety, harmony, and stability of society or relationships
Social Commitment	Preservation and enhancement of the welfare of all people
Stimulation	Excitement, novelty, and challenge in life

Table 1. Definitions of Schwartz's value model.

Even though Schwartz' theory has originally been developed as a theory of life values, it has also been widely used by other researchers to measure work values as well (e.g., Cable & Edwards, 2004; De Clerq et al., 2008). In the recent studies (De Clerq, 2006; De Clerq et al., 2008), it was demonstrated that Schwartz' value theory can provide a comprehensive framework to measure employees' work related values and their perceived organizational values. In our current study, we also used the work value survey adapted and tested by De Clerq and his colleagues (De Clerq et al., 2008). In their adapted version, they defined three bipolar factors that are summarized as:

1. Self-Transcendence (i.e., sacrificing selfish and individualistic concerns for the benefit of the organization and other colleagues) versus Self-Enhancement (i.e., enhancement of employee's own personal interests at the expense of the organization's and other colleagues' interests),
2. Openness to Change (i.e., following own intellectual and emotional interests in unpredictable and uncertain directions) versus Conservation (i.e., trying to preserve the status quo and preference of certainty in relationships), and
3. Goal-Orientedness (i.e., living and working to fulfill a purpose, not giving up) versus Hedonism (i.e., pleasure and sensuous gratification for oneself) (De Clerq, 2006; De Clerq et al., 2008).

In this framework, as a result of the bipolarity, consequences of any value are expected to have consequences that may conflict with their opposite value type. In other words, opposite consequences and associations are likely expected between two opposite values that constitute the poles of a bipolar factor and a third variable.

Value congruence, on the other hand, can be defined as the similarity between values held by individuals and organizations. Value congruence or the fit between employees' individual work values and perceived organizational values is widely accepted as supplementary person-organization (P-O) fit (Edwards & Cable, 2009; Kristof-Brown et al., 2005; Verquer et al., 2003). Research demonstrated that when employees' values fit the values of their organization, they are most likely satisfied and identified with and committed to their organization, and seek to maintain the employment relationship (Edwards & Cable, 2009; Amos & Weathington, 2008; Kristof-Brown et al., 2005; Verquer, Beehr, & Wagner, 2003). These consequences of value congruence are beneficial for both employees and organizations, as those outcomes foster extra-role behaviors linked to positive employee attitudes (Podsakoff et al., 2000).

Subjective fit captures employees' own perceptions about the extent to which they feel like they fit into their organization, whereas objective fit compares an employee's values with organizational values as seen by other people, such as managers or coworkers (Kristof-Brown et al., 2005). While both approaches seem feasible and are widely used, in our study, we conceptualize and assess value congruence using subjective fit, which entails the congruence between an employee's own values and his or her perceptions of the organization's values (Kristof-Brown et al., 2005). Because the conceptual aim of our study is to explain why P-O fit in terms of value congruence relates to employee attitudes, as TQM orientation. Employees' attitudes, such as TQM orientation, are subjective and therefore are expected to associate more strongly with the congruence between employee and organizational values as seen by the employee himself/herself than as seen by other members of the organization (Kristof-Brown et al., 2005; Edwards & Cable, 2009).

Another important issue for P-O fit studies, as recently emphasized by Kristof-Brown and colleagues (2005), is the necessity of having commensurate measures for supplementary P-O fit. Commensurability can be defined as describing both individual work values and organizational values with the same content dimensions (Kristof, 1996). Thus, in our study, we only include commensurate dimensions as work and organizational values.

3. Hypotheses development

Based on the arguments above, it is proposed that employees would be TQM oriented because certain individual work values may function as guiding principles in their organization and work environment. Coyle-Shapiro and Morrow (2003) suggested that individual differences might play an important role in predicting TQM orientation of the employees. In addition, recent research has also yielded that work and organizational values might be considered as an important source for employees' attitudes and behaviors (Edwards & Cable, 2009; Amos & Weathington, 2008). Therefore, we posit that there will be an association between the importance that employees attach to work related values and their TQM orientation. As values are fundamental properties of both people and organizations (see Cable & Edwards, 2004; Kristof-Brown et al., 2005), we also focus on their independent relationships with TQM orientation.

3.1 Self-transcendence / self-enhancement and TQM orientation

By definition, self-transcendence reflects the collectivist attitudes, which have substantial precedence of group goals over personal goals (Hofstede, 1980, 2001). In earlier studies, results have revealed that individuals, who hold collectivistic values or norms, are more likely expected to demonstrate collaborative and cooperative behaviors, and to perform in a team or work group environment, in which individual goals are subordinated to, and are aligned with the goals of the group (Wagner, 1995; Kirkman & Shapiro, 2001). Effective teamwork requires cooperation and sharing among individual employees and subordinating individual local interests to collective global interests. As such, employees' willingness or ability to embrace teamwork can be expected to depend on the extent to which they value self-transcendence. Employees, who hold self-transcendence as their value, greatly emphasize membership in a group and are prepared to look out for the well-being of the group even at the expense of their own personal interests (Wagner, 1995; Eby & Dobbins, 1997). On the other hand, research has also yielded that employees, who have precedence on self-enhancement, show greater resistance towards working in teams and negatively affect the performance of the work groups (Kiffin-Petersen & Cordery, 2003; Kirkman & Shapiro, 2001). Therefore, we hypothesize:

Hypothesis 1a: There will be a positive relationship between self-transcendence and teamwork.

Hypothesis 1b: There will be a negative relationship between self-enhancement and teamwork.

TQM has, by tradition, been defined as a collective approach using quality circles and empowered team-based methodologies to achieve continuous improvement (Dale, 1999). Continuous improvement, advocated by TQM, is based on members complementing and sharing their knowledge, working together and creating organizational learning for successful improvement initiatives (Harnesk & Abrahamsson, 2007). Thus, employees, who put emphasis on self-transcendence, are expected to put voluntarily more effort on collaborative team and workgroup activities, such as participating quality circles, and to show less resistance when it comes to apply changes throughout the organization for the well-being of the whole. On the other hand, self-enhancement or individualistic approach appears to be largely unworkable and incompatible with the original TQM's team-based philosophy and continuous improvement activities (McKenna & Beech, 2002). In other words, employees, who emphasize self-enhancement, are not expected to contribute to the improvement attempts, unless those attempts provide them with personal benefits. Therefore, we hypothesize:

Hypothesis 1c: There will be a positive relationship between self-transcendence and continuous improvement.

Hypothesis 1d: There will be a negative relationship between self-enhancement and continuous improvement.

3.2 Openness to change / conservation and TQM orientation

In our value framework, openness to change represents continuous attempt for creating and exploring new ideas and ways of doing business (De Clerq, 2006). A prerequisite for attaining development-oriented learning focused on improving quality for an organization's change initiatives is to have employees working in teams with a synergy for questioning established routines through alternative thinking (Ellström, 2001). Implementation of empowered, self-managing, innovative teams in TQM framework involves extensive methodological, structural and philosophical changes (Kirkman et al., 2000). For an employee to display a high performance in such a team environment, he/she needs to have a strong precedence on high tolerance for change. Sufficient evidence exists to suggest that an employee's receptivity to organizational change may be an important factor in their preference for working in a team (Kiffin-Petersen & Cordery, 2003). Hence, openness to change would most likely act as a buffer to reduce employee resistance and maintain their continued goodwill in the face of team-based forms of work organization that accompany the significant organizational changes. Therefore, we hypothesize:

Hypothesis 2a: There will be a positive relationship between openness to change and teamwork.

Hypothesis 2b: There will be a negative relationship between conservation and teamwork.

Openness to change assumes the ability to adopt new behaviors or strategies when old ones are no longer viable (Jerez-Gomez et al., 2005). Within the framework of TQM, embracing change, testing assumptions, shifting paradigms, tolerating ambiguity and taking risks, which are among the attitudes of employees open to change, are considered as crucial components for continuous improvement (Napolitano & Henderson, 1998; De Jager et al., 2004). In addition, commitment to change has been revealed as one of the most important and major motivations of individuals for introducing TQM, and one of the key elements for continuous improvement according to Kaizen (Dale, 1999). Employees, who emphasize openness to change as a value, are essential elements to create an organization capable of regenerating itself and coping with new challenges by leading the process of change through continuous improvement (Jerez-Gomez et al., 2005). On the other hand, in the cultures, in which employees has precedence on conservation, companies will not take avoidable risks, initiate continuous changes and adopt innovations if its effectiveness and value have already been proven (Waarts & van Everdingen, 2005). Thus, we hypothesize:

Hypothesis 2c: There will be a positive relationship between openness to change and continuous improvement.

Hypothesis 2d: There will be a negative relationship between conservation and continuous improvement.

3.3 Goal orientedness / hedonism and TQM orientation

Following the literature, a team can be defined as a social system of three or more people, which is embedded in an organization, whose members collaborate on a common goal (Salas, Cooke & Rosen, 2008). Teams are social entities composed of members with high task interdependency and shared and valued common goals (Salas et al., 2008). Teamwork is a set of interrelated and flexible cognitions, behaviors, and attitudes that are used to achieve desired mutual goals. Thus, as one can understand from the definitions, having common and shared goals is the core and fundamental element of being a productive and innovative team. Goal-oriented employees can be motivated towards the team goals and contribute to team success without giving up. In TQM framework, a defining basic feature of self-managing and empowered teams, for instance, is that team members contribute to a common goal and motivate themselves and each other to do so (Ellemers, Gilder & Halsam,

2004). Salas and his colleagues (2008) also suggest that self-managing teams containing employees, who value hard work and achievement, are expected to show higher levels of performance. On the other hand, hedonist employees are not expected to put effort to achieve teams' common goals, since they focus on their personal lives and needs. Thus, employees having hedonism as a value, are not considered as a good fit for teamwork. Hence, we hypothesize:

Hypothesis 3a: There will be a positive relationship between goal-orientedness and teamwork.

Hypothesis 3b: There will be a negative relationship between hedonism and teamwork.

Goal orientedness includes setting clear goals and results that provide targets for change, and opportunities to assess whether change has occurred during improvement initiatives. During continuous improvement activities, results of the changes are periodically measured and compared with the goals set, thus emphasis is on the goal accomplishment and achievement. According to the Juran (1989), Ishikawa (1985) and Dale (1999), setting goals for improvement is one of the fundamental elements of continuous improvement. In fact, setting goals is the first step for planning quality improvement. Hence, goal oriented employees are the key players for those continuous improvement initiatives with their persistence on achievement and reaching the goals. Dale (1999) also proposed that employees with high goal-orientedness usually encourage other employees to establish goals for improvement of organization and also to commit to the achievement of those goals. According to Juran (1989), certain roles of employees and especially top management include, among others, establishing and deploying quality goals and stimulating improvement. The European Quality Award and the Malcolm Baldrige Quality Award also recognize the crucial role of leadership in creating the goals and systems that guide the pursuit of continuous performance improvement (Dale, 1999). Thus, based on the arguments above, we expect a positive association between goal orientedness and continuous improvement, while a negative relationship is expected between hedonism and continuous improvement, as hedonism being at the opposite end of the continuum. Therefore, we hypothesize:

Hypothesis 3c: There will be a positive relationship between goal orientedness and continuous improvement.

Hypothesis 3d: There will be a negative relationship between hedonism and continuous improvement.

3.4 Value congruence and TQM orientation

One of the most important explanations of value congruence effects is that value congruence promotes communication within organizations (Erdogan, Kraimer & Liden, 2004). In the TQM context, communication refers to the open exchange of information through formal and informal interactions among organizational members that facilitates teamwork, continuous improvement and organizational learning (Edwards & Cable, 2009). In theory, value congruence foster teamwork, because having shared standards and values concerning what is important establishes a common frame for shared goals, as well as describing and interpreting events (Erdogan et al., 2004). This common understanding promotes harmony and collaboration among organizational and team members through the exchange of information. Having shared values, standards and understanding enhances the likelihood of success of continuous improvement initiatives by reducing the misunderstandings and conflicts among organizational members.

In addition, value congruence would increase predictability, which is defined as the confidence people have about how others will act and how events will unfold, because organizational members with shared values have similar motives, set similar goals, and respond to events in similar ways (O'Reilly et al., 1991; Edwards & Cable, 2009). Similarly, value congruence also promotes mutual understanding and reduces uncertainty through sharing knowledge and interpersonal communication (Berger, 2005). Uncertainty reduction effect of value congruence provides a basis for successful continuous improvement, which contains risk to a manageable extent and contributes to obtain more predictable outcomes.

Furthermore, recent research has also yielded that value congruence is positively related to organizational commitment and identification (Amos & Weathington, 2008; Edwards & Cable, 2009). Employees, who feel that their organization values the same things that they do, will likely have an emotional attachment to their organization. Organizational commitment and identification are important factors promoting teamwork, involvement, collaboration and extra-role behaviors (e.g., Van Knippenberg & Van Schie, 2000; Riketta, 2005). Employees, who identify strongly with their organizations, have a supportive attitude towards them. Thus, such employees would likely be willing to contribute to the organization through participation to teamwork, initiation of quality management activities and involvement to continuous improvement process.

Based on the above arguments, it is plausible to suggest that value congruence between employees and their employing organizations will have positive relationship with TQM orientation. Thus, we hypothesize:

Hypothesis 4a: There will be a positive relationship between value congruence and teamwork.

Hypothesis 4b: There will be a positive relationship between value congruence and continuous improvement.

4. Method

4.1 Sample and procedure

In order to test the proposed hypotheses empirically, the data were collected from white-collar employees working in 36 different organizations from Turkey, including banks, consultant firms, retail companies, public institutions and universities. Of the 520 employees surveyed, 219 employees provided usable responses, yielding a response rate of 42 percent.

All measurements included in the questionnaire were originally developed in English and translated into Turkish via the back-translation technique (see Brislin, 1980). Prior to administering the questionnaire, we conducted a pilot study, which revealed that scales were easily understood by white-collar employees targeted. Data was acquired via a structured questionnaire, which was distributed by our research assistant on site. Before handing out the questionnaires, our research assistant explained the purpose of the survey and noted that participation was completely voluntary. Furthermore, each questionnaire was also accompanied by a cover letter to ensure confidentiality. Respondents were asked to return the completed questionnaires directly to the research assistant to ensure their anonymity. No personal data was collected except demographics summarized below.

The average participant was 33 years old (standard deviation of 6.2 years). The participants were 68% male and 92% had at least a Bachelor's degree (university graduates). Participants had worked for their companies for an average of 6 years (standard deviation of 5.4 years).

4.2 Measurement of Work and Organizational Values (WOV)

Work and organizational values (WOV) were measured with a scale adapted from the 50-item Work and Organizational Values Survey (WOVS) developed and tested by De Clerq and his colleagues (2008) based on the value theory of Schwartz (1992). In this scale, there were 12 items measuring Self-Enhancement, 11 items for Self-Transcendence, 5 items for Openness to Change, 8 items for Conservation, 5 items for Hedonism and 9 items for Goal-Orientedness. A short explanatory statement was provided for each single value item (e.g., Perseverance: to carry on, not giving up; Self-indulgent: doing pleasant things). Importance of each single value was tapped on a 7-point scale from (-1): opposed to my / my organization's principles, through (0): not important for me / my organization, to (5): has supreme importance for me / my organization. This asymmetrical measurement scale was adopted from Schwartz (1992) and De Clerq et al. (2008). It mainly reflects to what extent those values are desirable and attractive. Participants rated each single value item based on its importance in their own work (personal work values) and the perceived importance of those values for the organization, in which they work (perceived organizational values).

The six values included in WOV measure, represent the opposite poles of three dimensions; namely self-enhancement versus self-transcendence, openness to change versus conservation, and hedonism versus goal-orientedness. The coefficient alpha reliabilities ranged from 0.79 to 0.91 for personal work values; and ranged from 0.72 to 0.86 for perceived organizational values.

4.3 Measurement of TQM orientation

TQM orientation was measured with a 15-item scale adapted from the scale developed and tested by Coyle-Shapiro and Morrow (2003). In this scale, there were 6 items measuring teamwork, 9 items for continuous improvement. As the result of factor analyses, three factors were extracted having eigenvalues greater than 1, which combined accounted for 63.9% of the total variance. The results yielded a single factor for teamwork as expected, yet two distinct factors for continuous improvement, namely active involvement and allegiance to quality, consistent with the previous study (Coyle-Shapiro & Morrow, 2003).

Teamwork represents to what extent employees value the performance of teamwork and exhibit a strong spirit of cooperation. Sample items of teamwork scale include "There is a lot of cooperation in my organization" and "The people in my organization encourage each other to work as a team". Coefficient alpha for this scale is 0.91. Active involvement represents to what extent an employee engages in quality focused behaviors. Sample items for active involvement scale include "I often put forward ideas and suggestions without expecting extra rewards" and "I put a lot of effort into thinking about how I can improve my work". On the other hand, allegiance to quality captures the extent of an employee's acceptance of quality and continuous improvement principles. A sample item for allegiance to quality is "Continuous improvement is essential for the future of my organization". Coefficient alpha reliabilities are 0.88 for active involvement and 0.86 for allegiance to quality.

4.4 Common method variance

The common method variance is a potential source of measurement error that may create a serious threat for the validity of conclusions about the associations among measures (Podsakoff et al., 2003). Because all data in this study are self-reported and collected through the same questionnaire during the same period of time with cross-sectional research design, the findings are not immune to common method variance, which may cause systematic measurement error and further bias the estimates of the true relationship among theoretical constructs (Podsakoff & Organ, 1986; Podsakoff et al., 2003).

Harman's one factor test and confirmatory factor analysis, and further post hoc statistical tests, were conducted to test the presence of common method effect. All six components of WOV (Self-Enhancement, Self-Transcendence, Openness to Change, Conservation, Hedonism and Goal-Orientedness) and three components of TQM orientation (Teamwork, Active Involvement and Allegiance to Quality) were entered into an exploratory factor analysis, using unrotated principal components factor analysis, principal component analysis with varimax rotation, and principal axis analysis with varimax rotation. This procedure was used to determine the number of factors necessary to account for the variance in the variables. If a substantial amount of common method variance is present, either (1) a single factor will emerge from the factor analysis, or (2) one general factor will account for the majority of the covariance among the variables (Podsakoff & Organ, 1986; Podsakoff et al., 2003).

The unrotated principal component factor analysis, principal component analysis with varimax rotation, and principal axis analysis with varimax rotation all revealed the presence of nine distinct factors with Eigenvalues greater than 1.0, rather than a single factor. Consistent with the expectation, all items loaded with high-standardized coefficients onto their respective factors and with substantially lower standardized coefficients in other factors. The nine factors together accounted for 67.7% of the total variance; the first (largest) factor did not account for a majority of the variance (12.6%). Thus, no general factor is apparent.

Moreover, all nine variables were loaded on one factor to examine the fit of the confirmatory factor analysis (CFA) model. If common method variance is largely responsible for the relationship among the variables, the one-factor CFA model should fit the data well. The confirmatory factor analysis showed that the single-factor model did not fit the data well; $\chi2=6309.86$, p<0.01, df=1952; $\chi2/df=3.23$ (>3); CFI=0.36, GFI=0.26, NNFI=0.28, RMSEA=0.16.

Although the results of the Harman's one-factor test and the single factor confirmatory factor analysis do not completely preclude the possibility of common method variance, they do suggest that common method variance is not of great concern and thus is unlikely to confound the interpretation of results for this study.

Constructs	Mean	SD	1	2	3	4	5	6	7	8	9	10	11	12	13	14	15
PWV																	
1. ST	2.23	0.84	0.80a														
2. SE	0.44	0.92	-0.34	0.91a													
3. OC	1.23	1.05	0.18	-0.06	0.89a												
4. CO	0.95	0.76	0.05	0.02	-0.38	0.81a											
5. GO	2.54	0.92	0.14	-0.11	0.09	-0.06	0.82a										
6. HE	-0.11	0.75	-0.09	0.13	-0.02	-0.03	-0.28	0.79a									
OWV																	
7. ST	2.54	0.92	0.45	-0.22	0.16	-0.09	0.19	-0.09	0.82a								
8. SE	0.61	0.88	-0.26	0.31	-0.02	0.06	-0.12	0.08	-0.19	0.85a							
9. OC	0.94	0.90	0.11	-0.01	0.33	-0.14	0.06	-0.07	0.11	-0.09	0.86a						
10. CO	0.25	0.82	-0.02	0.05	-0.08	0.39	-0.01	-0.02	-0.08	0.01	-0.20	0.77a					
11. GO	3.38	1.12	0.12	-0.07	0.10	-0.01	0.41	-0.12	0.12	-0.11	0.15	-0.04	0.80a				
12. HE	-0.25	0.77	-0.06	0.03	-0.03	0.00	-0.16	0.29	-0.13	0.08	-0.04	0.02	-0.24	0.72a			
TQM																	
13. TW	4.12	0.52	0.36	-0.16	0.19	-0.10	0.27	-0.08	0.39	-0.19	0.17	-0.06	0.23	-0.05	0.91a		
14. AI	3.93	0.65	0.28	-0.08	0.16	-0.03	0.15	-0.11	0.31	-0.14	0.15	-0.06	0.27	-0.07	0.39	0.88a	
15. AQ	3.78	0.72	0.17	-0.03	0.18	-0.07	0.09	-0.04	0.20	-0.13	0.18	-0.02	0.20	-0.04	0.28	0.36	0.86a

Correlations greater than or equal to 0.07 are statistically significant at p<0.05 level.
a Cronbach alpha reliability

PWV: Personal Work Values,   OWV: Organizational Work Values,   TQM: TQM Orientation
ST: Self transcendence,   SE: Self-enhancement,   OC: Openness to change,
CO: Conservation,   GO: Goal orientedness,   HE: Hedonism,
TW: Teamwork,   AI: Active Involvement,   AQ: Allegiance to quality

Table 2. Descriptive statistics and correlations among measures.

4.5 Factor analyses

In order to confirm the validity and reliability of factor structures, both exploratory (using SPSS 13.0) and confirmatory factor analyses (using AMOS 7.0) were performed on the sample. Prior to the estimation of the confirmatory measurement model, exploratory factor analyses were conducted to assess unidimensionality. In each of these analyses, a single factor was extracted (using a cut-off point of eigenvalue = 1), suggesting that our measurement scales were unidimensional. Next, consistent with our measurement theory, 50 items measuring WOV were hypothesized to load on six distinct factors (i.e.; Self-Enhancement, Self-Transcendence, Openness to Change, Conservation, Hedonism and Goal-Orientedness) in the measurement model. In addition, the 6 items measuring teamwork, 5 items measuring active involvement and 4 items measuring allegiance to quality were averaged to create composite indicants for each of these formative measures, which are then posited to load on three distinct factors in the measurement model. Alpha reliabilities for the scales ranged from 0.72 to 0.91 and factor loadings of items varied from 0.44 to 0.86. Means, standard deviations, alpha reliabilities and intercorrelations among the established measures are depicted in Table 2.

Next, confirmatory factor analysis was estimated on 65 items measuring nine constructs. In addition, it was also checked the measurement properties of the variables by comparing the baseline model with alternate models. Suggested nine-factor model resulted in a significant chi-square statistic and goodness-of-fit indices suggesting that the model fits the observed covariances well ($\chi 2=3467.77$, $p<0.01$, $df=1854$, $\chi 2/df=1.87$ (<3), CFI=0.95; GFI=0.90; NNFI=0.88; RMSEA=0.06). In addition, all items loaded significantly on their respective constructs (with the lowest t-value being 3.58), providing support for the convergent validity of measurement items. Finally, discriminant validity was obtained for all constructs since the variance extracted for each construct was greater than its squared correlations with other constructs (Fornell & Larcker, 1981).

4.6 Hypotheses analysis

Hypotheses relating values and value congruence to TQM orientation were tested using polynomial regression analysis (Edwards, 1994, 2002). Most research on the congruence between two constructs as a predictor of outcomes uses difference scores. Nevertheless, the main problem with difference scores is that the independent contribution of personal variables and organizational variables is ignored. Because a difference score is a calculated measure, it captures nothing more than the combined effects of its components. Investigating the issue without considering the measures separately as personal and organizational variables, it is difficult to address the question of whether outcome attitude or behavior is determined by personal characteristics, organizational characteristics, or their congruence (De Clerq, 2006). There may be mean-level differences when congruence occurs on a high level of the predictors versus a low level, or the incongruence in one direction (i.e., predictor X > predictor Y) may have different effects than the incongruence in the opposite direction (X < Y) (Edwards, 2002). Therefore, polynomial regression analysis and three-dimensional surface plots (surface response analysis) recommended by Edwards (1994, 2002) are among the best statistical techniques to more precisely examine the exact nature of personal-organizational work value congruence on employees' TQM orientation.

Constructs	Step 1			Step 2						
	X	Y	R²	X	Y	X²	XY	Y²	R²	ΔR²
Dependent Variable: Teamwork										
Self Transcendence	0.36**	0.44**	0.112**	0.28**	0.39**	0.05	-0.12*	0.03	0.134**	0.022
Self Enhancement	-0.11*	-0.13*	0.049*	-0.06	-0.08	0.06	-0.02	0.04	0.058	0.009
Openness to change	0.12*	0.11*	0.038*	0.10	0.09	0.11	-0.04	0.05	0.045	0.007
Conservation	-0.04	-0.02	0.016	-0.04	-0.03	0.01	-0.01	0.02	0.018	0.002
Goal Orientedness	0.24**	0.18**	0.086*	0.22**	0.14*	0.05	-0.04*	0.04	0.097*	0.011
Hedonism	-0.08	-0.06	0.022	-0.06	-0.04	0.04	-0.05	0.03	0.025	0.003
Dependent Variable: Active Involvement										
Self Transcendence	0.22**	0.29**	0.083**	0.17**	0.22**	0.06	-0.06*	0.04	0.098**	0.015
Self Enhancement	-0.08	-0.09	0.035	-0.05	-0.07	0.02	-0.02	0.03	0.039	0.004
Openness to change	0.07	0.07	0.026	0.05	0.06	0.03	-0.03	0.05	0.029	0.003
Conservation	-0.03	-0.02	0.019	-0.02	-0.02	0.01	-0.01	0.02	0.020	0.001
Goal Orientedness	0.18**	0.14*	0.081*	0.18*	0.12*	0.07	-0.02*	0.02	0.089*	0.008
Hedonism	-0.10*	-0.09*	0.031*	-0.09	-0.07	0.04	-0.02	0.03	0.033	0.002
Dependent Variable: Allegiance to quality										
Self Transcendence	0.15*	0.20**	0.078*	0.11*	0.16**	0.02	-0.03*	0.02	0.094**	0.016
Self Enhancement	-0.07	-0.09	0.023	-0.04	-0.05	0.06	-0.04	0.01	0.029	0.006
Openness to change	0.10*	0.09*	0.041*	0.10	0.08	0.07	-0.03	0.03	0.048	0.007
Conservation	-0.03	-0.01	0.008	-0.02	-0.01	0.01	-0.02	0.01	0.012	0.004
Goal Orientedness	0.16*	0.10*	0.065*	0.15*	0.12*	0.03	-0.04*	0.03	0.075*	0.010
Hedonism	-0.06	-0.05	0.022	-0.04	-0.03	0.04	-0.05	0.02	0.025	0.003

** $p<0.01$ * $p<0.05$ X: Personal Work Values   Y: Organizational Work Values

- For all columns except R² and ΔR², figures are unstandardized regression coefficients with all predictors entered simultaneously.
- For Model 1, the column labeled R² indicates the variance explained by two predictors (X, Y); for Model 2, the column R² indicates the variance explained by five predictors (X, Y, X², XY, Y²). The column labeled ΔR² contains incremental variance explained by the quadratic terms (X², Y²) and the congruence term (XY) over Model 1.

Table 3. Results of linear and quadratic regression analyses.

In order to be able to overcome aforementioned methodological problems associated with measures of congruence, we employed polynomial regression and surface response analysis to explore the hypothesized associations between personal and organizational work values and TQM orientation. It comprises a collection of procedures for estimating and interpreting three-dimensional surfaces relating two variables to an outcome.

Before applying the supplementary person-organization fit (work value congruence) model, employee's personal work values (X) and organizational work values (Y) were simultaneously entered to the equation to test their linear main effects in Step 1. Following this, to investigate supplementary person-organization fit, polynomial regression analysis was applied as Step 2 to represent the relationship between the congruence of employees' personal and perceived organizational values and their TQM orientation. In Step 2, personal work values squared (X^2), the interaction between personal work values and organizational work values (XY), and organizational work values squared (Y^2) were entered to the equation. To reduce multicollinearity and facilitate interpretation of the graphs, all predictor variables were scale-centered as suggested by previous research (Edwards, 1994; Shanock et al., 2010). A separate polynomial regression was run and examined for each of the six work values, which made six different polynomial regression analyses. The following regression equations were used to determine whether research hypotheses were supported. Results of the regression analyses are presented in Table 3.

$$\text{TQM Orientation} = b_0 + b_1X + b_2Y + e \qquad \text{(Step 1)}$$

$$\text{TQM Orientation} = b_0 + b_1X + b_2Y + b_3X^2 + b_4XY + b_5Y^2 + e \qquad \text{(Step 2)}$$

4.7 Results

Our first hypothesis suggested that self transcendence will be positively related to TQM orientation, whereas there will be a negative relationship between self enhancement and TQM orientation. First of all, proposed regression models in Step 1 consisting of the associations between self transcendence and TQM orientation components were found to be statistically significant. Regression analysis revealed that both personal self transcendence and perceived organizational self transcendence are positively related to all three dimensions of TQM orientation. It was also revealed that, influence of perceived organizational self transcendence on TQM orientation is stronger than personal self transcendence. Thus, our hypotheses 1a and 1c are fully supported. However, hypothesized negative relationship between personal and organizational self enhancement and TQM orientation was found to be statistically significant only for teamwork. Similarly, negative relationship between organizational self enhancement and teamwork was stronger than the relationship of personal self enhancement. So, our hypothesis 1b was supported, whereas 1d was not supported.

Our second hypothesis, on the other hand, suggested that openness to change will be positively related to TQM orientation, whereas a negative relationship between self conservatism and TQM orientation is expected. Regression equations between the openness to change and teamwork, as well as allegiance to quality, were found to be statistically significant, whereas the equation between openness to change and active involvement was not significant. Consequently, analysis confirmed the expected positive associations of openness to change with teamwork and allegiance to quality, but did not corroborate its positive relation with active involvement. Contrary to the influence of self transcendence, personal openness to change was found to have stronger association with TQM orientation than organizational openness to change. Thus, our hypothesis 2a was fully supported, while 2c was partially supported. For the relationship between conservation and TQM orientation, none of the regression equations were found to be statistically significant. Accordingly, hypothesized negative relationships between conservatism and TQM orientation were not statistically significant. Thus, our hypotheses 2b and 2d were not supported.

Third hypothesis suggested that goal orientedness will be positively associated with TQM orientation, while hedonism will be negatively related to TQM orientation. Regression analyses revealed that all of the proposed positive relationships between goal orientedness and TQM orientation components were statistically significant. Analyses also show that personal goal orientedness had a stronger influence on TQM orientation than organizational goal orientedness. Thus, our hypotheses 3a and 3c were fully supported. On the contrary, from the hypothesized negative relationships between hedonism and TQM orientation, only the one between hedonism and active involvement was found to be statistically significant, in which the negative influence of personal and organizational hedonism on TQM orientation were at the same extent. Thus, our hypothesis 3b was not supported, yet 3d was partially supported.

Finally, our fourth hypothesis proposed that the congruence between the personal values and the perceived organizational values will have a positive influence on TQM orientation. Polynomial regression analyses revealed that the quadratic equations were statistically significant for only two values, which are self transcendence and goal orientedness. For self transcendence and goal orientedness, Step 2 significantly increased the amount of variance accounted for over Step 1 in all three equations having each of three of TQM orientation components as dependent variable. Thus, we can conclude that our hypotheses 4a and 4b were partially supported.

To better comprehend the differentiated effects of personal and organizational values, as well as of their congruence, surface response analyses were also conducted. Three dimensional (3-D) graphs of the surface response analyses are displayed in Figure 1 and Figure 2.

Examining the surface plots for self transcendence (Figure 1) and goal orientedness (Figure 2), we found that all of the three TQM orientation components; namely teamwork, active involvement and allegiance to quality; were at the highest level when those values (self transcendence and goal orientedness) were rated as highly important both for the employee and the organization (shown by the highest point in the far back corner of graphics). Contrarily, all three TQM orientation components were lowest when the values were rated as not important both for the employee and the organization (shown by the lowest point in the near front corner of graphics).

In addition to that, in case of incongruence of the importance of the self transcendence for the employee and for the organization, organizational values have slightly stronger effect on TQM orientation than personal values for self transcendence. In case of incongruence for goal orientedness, the relationship showed a different pattern. For goal orientedness, it was revealed that personal values have slightly stronger effect on TQM orientation than organizational values.

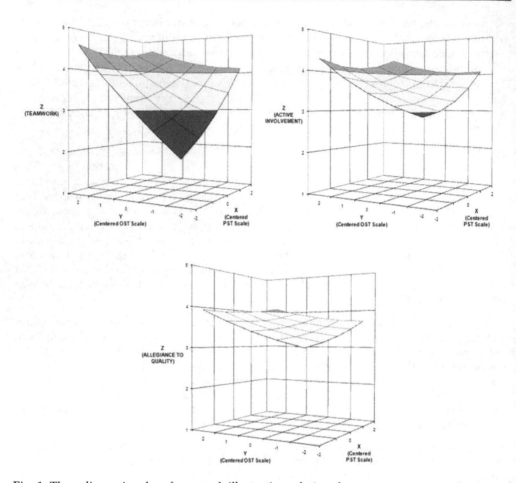

Fig. 1. Three-dimensional surface graph illustrating relations between person-organization fit for self transcendence and TQM orientation. **(OST: Organizational Self Transcendence, PST: Personal Self Transcendence)**

However, it was also yielded that, when those two values were rated as important for the employee but not important for the organization, or when they were rated as important for the organization and not important for the employee (in case of incongruence), TQM orientation were still high (see far left and right corners of the graphics). This pattern of relationships suggests that "exact congruence" between the importance attached to self transcendence and goal orientedness by the employee and perceived importance of those values for the organization is not substantially necessary for positive TQM orientation.

While examining the relationships between personal – organizational value congruence and the three components of TQM orientation, polynomial regression analyses revealed that the combinations of personal – organizational values were only significant for two of the six values (self transcendence and goal orientedness). Since quadratic regression equations calculated for the personal – organizational congruence of the other values' yielded that the unstandardized regression coefficients in those equations are not statistically significant, their graphical illustrations are not depicted.

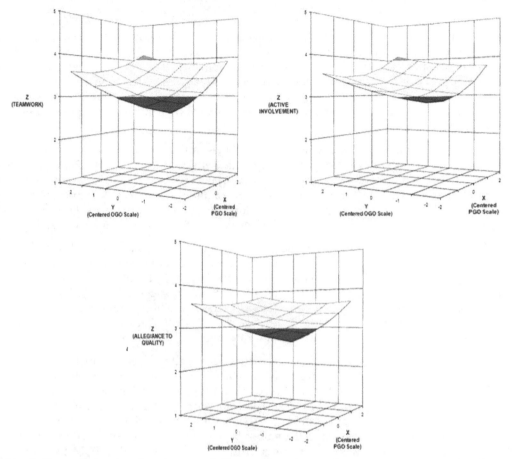

Fig. 2. Three-dimensional surface graph illustrating relations between person-organization fit for goal orientedness and TQM orientation. **(OGO: Organizational Goal Orientedness, PGO: Personal Goal Orientedness)**

5. Discussion and conclusion

In general, this present research was conducted to examine the associations between supplementary person-organization fit in terms of personal – organizational value congruence and components of employees' TQM orientation, which are participation in teamwork and involvement in continuous improvement. In addition, direct linear relationships between six values, which are self transcendence, self enhancement, openness to change, conservatism, goal orientedness and hedonism, and three TQM orientation components were also investigated.

Proposed relationships were studied by means of a commensurate and comprehensive work and organizational values survey. All respondents made judgments about their own personal work values and provided their perception about the importance of these same values for their organization.

First, significant linear effects were also yielded. The results of the two step polynomial regression analyses supported the hypothesized positive links from self transcendence and goal orientedness to all of the three TQM orientation components, which are teamwork, active involvement and allegiance to quality, for both personal and organizational level; whereas openness to change was found to be positively associated with teamwork and allegiance to quality. On the other hand, only the negative relationships between self enhancement and teamwork and between hedonism and active involvement were found to be statistically significant.

Results suggest that the values, which were hypothesized as having positive relationships with TQM orientation, have stronger and significant effect on the outcomes. However, the values at the opposite poles, which were hypothesized as having negative associations with the outcome variables, were not revealed to have significant influence as expected. Thus, one can conclude that the attitudes and behaviors, which do not promote the benefit of the organization, are not substantially affected by the values of the people or the organization.

Furthermore, curvilinear relationship between the value congruence and TQM orientation was also investigated through polynomial regression and surface response analyses. Results yielded that that the quadratic equations were statistically significant for only two values, which are self transcendence and goal orientedness. Findings suggest that, TQM orientation would be at the highest level when the values are perceived as highly important both for the employee and the organization. On the contrary, TQM orientation would be lowest when the values were rated as not important both for the employee and the organization.

More importantly, even in case of inconsistency between the level of importance of the values for employees and the organization; in other words, even when values were rated as important for the employee but not important for the organization, or when they were rated as important for the organization and not important for the employee, TQM orientation would still be at a high level. This finding suggest that "exact congruence" between the importance attached to the values by the employee and perceived importance of those values for the organization is not crucial to obtain a positive TQM orientation among employees.

Besides the aforementioned theoretical implications of the results, this study also offers several practical implications for practicing managers. Based on the positive relations between value congruence and positive work related attitudes, such as TQM orientation, managers should attempt to foster value congruence to improve the attitudes and behaviors of employees, which would be beneficial for the organization. Thus, managers had better devote more energy and resources assessing value congruence when hiring job applicants (Cable & Judge, 1996) and engage in socialization tactics to adjust the values of employees in the course of the values, which are perceived as important throughout the organization (Edwards & Cable, 2009). In addition, to foster the positive consequences of person – organization fit, managers may also explore strategies that may directly influence the facilitators of the value congruence, such as trust and communication.

Although it was revealed that value congruence had an important role in practical managerial aspects, findings of this study also suggested that some values have considerable direct linear effect on employees' attitudes and behaviors, even in case of inconsistency between the importance of the values for the employee and perceived importance of them in the organization. Hence, employees would also rather to explore those values that may have a positive impact for the organization, and try to find ways to promote their importance among employees throughout the organization.

In conclusion, this study's findings suggest that, in addition to the linear effects of values that explain a considerable variance, value congruence effects also play an important role in explaining the adoption of TQM orientation. In addition, this paper highlights the importance of using value dimensions and demonstrates that different value types might have different relationships with various components of quality orientation.

6. Limitations and recommendations for future research

The findings and the contribution of the current study must be evaluated taking into account the potential limitations of the research design. First, the data was cross-sectional, making it impossible to imply causality. All of the variables were measured at the same time and from the same source, so concern over the effects of common method variance was warranted. To minimize this potential problem, the scales in the actual survey were ordered so that the dependent variable did not precede all the independent ones (Podsakoff & Organ, 1986). In addition, as explained in detail in the relevant section above, the results of the Harman's one-factor test and the single factor confirmatory factor analysis suggest that common method variance is not of great concern and thus is unlikely to confound the interpretation of results. Furthermore, variance inflation factors and condition indexes, which were calculated well-below the threshold values, indicates no problem regarding a common-method bias.

Second, any data collected by self-report measures may have been influenced by a social desirability response bias. Although one cannot rule out a self-serving bias as a possible influence, researchers have suggested that social desirability is generally not a source of bias in measuring organizational perceptions (Podsakoff & Organ, 1986).

Finally, our sample comprised of employees working in different public and private organizations in Turkey. This setting and findings may not be generalized to all employees and organizations. Hence, implications and conclusions of this study are bounded by the context of the research, but future research could involve the replication of this study in a number of different contexts. The author believes that future research assessing similar data from different contexts will provide informative validation for the results of this study. Additionally investigating other firm-specific effects and managerial implications that may promote value congruence and person – organization fit may guide academicians and practitioners to better understand the determinants, consequences and the benefits of the notion for fit.

Although the limitations mentioned above, this paper provides important theoretical implications for P-O fit theory and contributes to the literature in terms of TQM and human resource management perspectives. Furthermore, findings also offer several insights for practitioners. The results may assist managers to make better decisions in opting for an appropriate management scheme in order to achieve better TQM performance.

7. References

Amos, E.A. & Weathington, B.L. (2008). An Analysis of the Relation Between Employee-Organization Value Congruence and Employee Attitudes. *Journal of Psychology*, Vol.142, No.6, pp.615-631.

Berger, C.R. (2005). Interpersonal Communication: Theoretical Perspectives, Future Prospects. *Journal of Communication*, Vol.55, pp.415–447.

Bretz, R.D. & Judge, T.A. (1994). Person-Organization Fit and the Theory of Work Adjustment: Implications for Satisfaction, Tenure, and Career Success. *Journal of Vocational Behavior*, Vol.44, pp.32-54.

Brislin, R.W. (1980). Translation and Content Analysis of Oral and Written Materials. In: *Handbook of Cross-Cultural Psychology*, H. C. Triandis & J. W. Berry (Eds.), Vol. 2, pp. 389–444, Boston: Allyn & Bacon.

Cable, D.M. & Judge, T.A. (1996). Person-Organization Fit, Job Choice Decisions, and Organizational Entry. *Organizational Behavior and Human Decision Processes*, Vol.67, No.3, pp.294-311.

Caplan, R.D. (1987). Person-Environment Fit Theory and Organizations: Commensurate Dimensions, Time Perspectives, and Mechanisms. *Journal of Vocational Behavior*, Vol.31, pp.248-267.

Chatman, J.A. (1989). Improving Interactional Organizational Research: A Model of Person–Organization Fit. *Academy of Management Review*, Vol.14, pp.333–349.

Choi, T.Y. & Behling, O.C. (1997). Top Managers and TQM Success: One More Look After All These Years. *Academy of Management Executive*, Vol.11, pp.37-47.

Coyle-Shapiro, J. & Morrow, P. (2003). The Role of Individual Differences in Employee Adoption of TQM Orientation. *Journal of Vocational Behavior*, Vol.62, pp.320-340.

Dale, B.G. (1999). *Managing Quality, 3rd ed.* Oxford: Blackwell.

De Clerq, S. (2006). *Extending the Schwartz Value Theory for Assessing Supplementary Person-Organization Fit*. Unpublished Doctoral Dissertation, Gent University, Gent, Belgium.

De Clerq, S., Fontaine, J.R.J. & Anseel, F. (2008). In Search of a Comprehensive Value Model for Assessing Supplementary Person-Organization Fit. *Journal of Psychology*, Vol.142, No.3, pp.277-302.

De Jager, B., de Jager, J., Minnie, C., Welgemod, M., Bessant. J. & Francis. D. (2004). Enabling Continuous Improvement: A Case Study of Implementation. *Journal of Manufacturing Technology Management*, Vol.15, No.4, pp.315-317.

Deming, W.E. (1986). *Out of the Crisis*. Cambridge, MA: MIT Press.

Eby, L.T. & Dobbins, G.H. (1997). Collectivist Orientation in Teams: An Individual and Group-Level Analysis. *Journal of Organizational Behavior*, Vol.18, pp.275-295.

Edwards, J.R. (1994). Regression Analysis as an Alternative to Difference Scores. *Journal of Management*, Vol.20, pp.683-689.

Edwards, J.R. (2002). Alternatives to Difference Scores: Polynomial Regression Analysis and Response Surface Methodology. In: *Measuring and Analyzing Behavior in Organizations: Advances in Measurement and Data Analysis*, F.Dragow & N.Schmitt (Eds.), pp. 350-400. San Francisco, CA: Jossey-Bass.

Edwards, J.R. & Cable, D.M. (2009). The Value of Value Congruence. *Journal of Applied Psychology*, Vol.94, No.3, pp.654-677.

Ellemers, N., Gilder, D. & Haslam, S.A. (2004). Motivating Individuals and Groups at Work: A Social Identity Perspective on Leadership and Group Performance. *Academy of Management Review*, Vol.29, No.3, pp.459-478.

Ellström, P.E. (2001). Integrating Learning and Work: Conceptual Issues and Critical Conditions. *Human Resource Development Quarterly*, Vol. 12 No. 4, pp. 421-435.

Erdogan, B., Kraimer, M.L. & Liden, R.C. (2004). Work Value Congruence and Intrinsic Career Success: The Compensatory Roles of Leader-Member Exchange and Perceived Organizational Support. *Personnel Psychology*, Vol.57, pp.305-332.

Fornell, C. & Larcker, D. (1981). Evaluating Structural Equation Models with Unobservable Variables and Measurement Error. *Journal of Marketing Research*, Vol.18, pp.39–50.

Hackman, R. & Wageman, R. (1995). Total Quality Management: Empirical, Conceptual, and Practical Issues. *Administrative Science Quarterly*, Vol.40, pp.309-342.

Harnesk, R. & Abrahamsson, L. (2007). TQM: An Act of Balance between Contradictions. *The TQM Magazine*, Vol.19, No.6, pp.531-540.

Hofstede, G. (1980). Motivation, Leadership and Organization: Do American Theories Apply Abroad?. *Organizational Dynamics*, Vol.9, pp.42–63.

Hofstede, G. (2001). *Culture's Consequences: Comparing Values, Behaviors, Institutions, and Organizations across Nations*. New York: Sage Publications.

Ishikawa, K. (1985). *What is Total Quality Control? The Japanese Way*. Englewood Cliffs, N.J.: Prentice Hall.

Jerez-Gomez, P., Cespedes-Lorente, J.& Valle-Cabera, R. (2005). Organizational Learning Capability: A Proposal of Measurement. *Journal of Business Research*, Vol.58, pp.715-725.

Juran, J. M. (1989). Universal Approach to Managing for Quality. *Executive Excellence*, Vol.6, No.5, pp.15-17.

Kiffin-Petersen, S.A. & Cordery, J.L. (2003). Trust, Individualism and Job Characteristics as Predictors of Employee Preference for Teamwork. *The International Journal of Human Resource Management*, Vol.14, No.1, pp.93-116.

Kirkman. B.L., Jones. R.G. & Shapiro. D.L. (2000). Why do Employees Resist Teams? Examining the 'Resistance Barrier' to Work Team Effectiveness. *International Journal of Conflict Management*, Vol.11, pp.74-92.

Kirkman, B.L. & Shapiro, D.L. (2001). The Impact of Cultural Values on Job Satisfaction and Organizational Commitment in Self-Managing Work Teams: The Mediating Role of Employee Resistance. *Academy of Management Journal*, Vol.44, pp.557-569.

Kristof, A.L. (1996). Person-Organization Fit: An Integrative Review of its Conceptualizations, Measurement, and Implications. *Personnel Psychology*, Vol.49, No.1, pp.1-49.

Kristof-Brown, A.L., Zimmerman, R.D. & Johnson, E.C. (2005). Consequences of Individuals' Fit at Work: A Meta Analysis of Person-Job, Person-Organization, Person-Group and Person-Supervisor Fit. *Personnel Psychology*, Vol.58, pp.281-342.

Muchinsky, P.M. & Monahan, C.J. (1987). What is Person-Environment Congruence? Supplementary versus Complementary Models of Fit. Journal of Vocational Behavior, Vol.31, pp.268-277.

Napolitano, C.S. & Henderson, L.J. (1998). *The Leadership Odyssey: A Self Development Guide to New Skills for New Times*. San Francisco: Jossey-Bass.

O'Reilly, C.A., Chatman, J. & Caldwell, D.F. (1991). People and Organizational Culture: A Profile Comparison Approach to Assessing Person-Organization Fit. *Academy of Management Journal*, Vol.34, No.3, pp.487-516.

Ostroff, C., Shin, Y. & Kinicki, A.J. (2005). Multiple Perspectives of Congruence: Relationships between Value Congruence and Employee Attitudes. *Journal of Organizational Behavior*, Vol.26, pp.591-623.

Podsakoff, P.M. & Organ, D.W. (1986). Self-Reports in Organizational Research: Problems and Prospects. *Journal of Management*, Vol.12, pp.531-544.

Podsakoff, P.M., MacKenzie, S.B., Paine, J.B. & Bachrach, D.G. (2000). Organizational Citizenship Behaviors: A Critical Review of the Theoretical and Empirical Literature and Suggestions for Future Research. *Journal of Management*, Vol.26, pp.513–563.

Podsakoff, P.M., MacKenzie, S.B., Lee, J.Y. & Podsakoff, N.P. (2003). Common Method Biases in Behavioral Research: A Critical Review of the Literature and Recommended Remedies. *Journal of Applied Psychology*, Vol.88, pp.879-903.

Posner, B.Z. (1992). Person-Organization Values Congruence: No Support for Individual Differences as a Moderating Influence. *Human Relations*, Vol.45, pp.351-361.

Reger, R.K., Gustafson, L.T., DeMarie, S.M. & Mullane, J.V. (1994). Reframing the Organization: Why Implementing Total Quality is Easier Said than Done. *Academy of Management Review*, Vol.19, pp.565-584.

Riketta, M. (2005). Organizational Identification: A Meta-Analysis. *Journal of Vocational Behavior*, Vol.66, pp.358–384.

Saks, A.M. & Ashforth, B.E. (2002). Is Job Search Related to Employment Quality? It All Depends on the Fit. *Journal of Applied Psychology*, Vol.87, No.4, pp.646-654.

Salas, E., Cooke, N.J. & Rosen, M.A. (2008). On Teams, Teamwork, and Team Performance: Discoveries and Developments. *Human Factors: The Journal of Human Factors and Ergonomics Society*, Vol.50, pp.540-547.

Schwartz, S.H. (1992). Universals in the Content and Structure of Values: Theoretical Advances and Empirical Tests in 20 Countries. *Advances in Experimental Social Psychology*, Vol.25, pp.1–65.

Schwartz, S.H., Melech, G., Lehmann, A., Burgess, S., Harris, M. & Owens, V. (2001). Extending the Cross-Cultural Validity of the Theory of Basic Human Values with a Different Method of Measurement. *Journal of Cross-Cultural Psychology*, Vol.32, pp.519-542.

Sitkin, S.B., Sutcliffe, K.M. & Schroeder, R.G. (1994). Distinguishing Control from Learning in Total Quality Management: A Contingency Perspective. *Academy of Management Review*, Vol.19, pp.537-564.

Ucanok, B. (2009). The Effects of Work Values, Work-Value Congruence and Work Centrality on Organizational Citizenship Behavior. *International Journal of Behavioral, Cognitive, Educational and Psychological Sciences*, Vol.1, No.1, pp.1-14.

Van Knippenberg, D. & Van Schie, E.C.M. (2000). Foci and Correlates of Organizational Identification. *Journal of Occupational and Organizational Psychology*, Vol.73, pp.137-148.

Verquer, M.L., Beehr, T.A. & Wagner, S.H. (2003). A Meta-Analysis of Relations Between Person-Organization Fit and Work Attitudes. *Journal of Vocational Behavior*, Vol.63, pp.473-489.

Waarts, E. & van Everdingen, Y.M. (2005). The Influence of National Culture on the Adaptation Status of Innovations: An Empirical Study of Firms Across Europe. *European Management Journal*, Vol.23, No.6, pp.601–610.

Wagner, J.A. (1995). Studies of Individualism-Collectivism: Effects on Cooperation in Groups. *Academy of Management Journal*, Vol.38, pp.152-172.

Permissions

The contributors of this book come from diverse backgrounds, making this book a truly international effort. This book will bring forth new frontiers with its revolutionizing research information and detailed analysis of the nascent developments around the world.

We would like to thank Dr. Kim-Soon Ng, for lending his expertise to make the book truly unique. He has played a crucial role in the development of this book. Without his invaluable contribution this book wouldn't have been possible. He has made vital efforts to compile up to date information on the varied aspects of this subject to make this book a valuable addition to the collection of many professionals and students.

This book was conceptualized with the vision of imparting up-to-date information and advanced data in this field. To ensure the same, a matchless editorial board was set up. Every individual on the board went through rigorous rounds of assessment to prove their worth. After which they invested a large part of their time researching and compiling the most relevant data for our readers. Conferences and sessions were held from time to time between the editorial board and the contributing authors to present the data in the most comprehensible form. The editorial team has worked tirelessly to provide valuable and valid information to help people across the globe.

Every chapter published in this book has been scrutinized by our experts. Their significance has been extensively debated. The topics covered herein carry significant findings which will fuel the growth of the discipline. They may even be implemented as practical applications or may be referred to as a beginning point for another development. Chapters in this book were first published by InTech; hereby published with permission under the Creative Commons Attribution License or equivalent.

The editorial board has been involved in producing this book since its inception. They have spent rigorous hours researching and exploring the diverse topics which have resulted in the successful publishing of this book. They have passed on their knowledge of decades through this book. To expedite this challenging task, the publisher supported the team at every step. A small team of assistant editors was also appointed to further simplify the editing procedure and attain best results for the readers.

Our editorial team has been hand-picked from every corner of the world. Their multi-ethnicity adds dynamic inputs to the discussions which result in innovative outcomes. These outcomes are then further discussed with the researchers and contributors who give their valuable feedback and opinion regarding the same. The feedback is then collaborated with the researches and they are edited in a comprehensive manner to aid the understanding of the subject.

Apart from the editorial board, the designing team has also invested a significant amount of their time in understanding the subject and creating the most relevant covers. They scrutinized every image to scout for the most suitable representation of the subject and create an appropriate cover for the book.

The publishing team has been involved in this book since its early stages. They were actively engaged in every process, be it collecting the data, connecting with the contributors or procuring relevant information. The team has been an ardent support to the editorial, designing and production team. Their endless efforts to recruit the best for this project, has resulted in the accomplishment of this book. They are a veteran in the field of academics and their pool of knowledge is as vast as their experience in printing. Their expertise and guidance has proved useful at every step. Their uncompromising quality standards have made this book an exceptional effort. Their encouragement from time to time has been an inspiration for everyone.

The publisher and the editorial board hope that this book will prove to be a valuable piece of knowledge for researchers, students, practitioners and scholars across the globe.

List of Contributors

Ng Kim-Soon
Universiti Tun Hussein, Onn Malaysia, Malaysia

Simmy Marwa
Portsmouth Business School, University of Portsmouth, Portsmouth, UK

Luminita Gabriela Popescu
National School of Political Studies and Public Administration, Romania

Soleyman Iranzadeh
Department of Management, Tabriz Branch, Islamic Azad University, Tabriz, Iran

Farzam Chakherlouy
Tabriz Business Training Center, East Azerbaijan, Tabriz, Iran

Luděk Šprongl
Central Laboratory, Šumperk Hospital, Czech Republic

Ali M. Al-Shehri
King Saud bin Abdulaziz University for Health Sciences, National Guard Health Affairs, Kingdom of Saudi Arabia

Wieslaw Urban
Bialystok University of Technology, Faculty of Management, Poland

Shu-Mei Wang
Department of Tourism, Shih Hsin University, Taipei, Taiwan

Jui-Min Li
Central Taiwan University of Science and Technology, Taichung City, Taiwan, Republic of China

Zulnaidi Yaacob
Universiti Sains Malaysia, Malaysia

Mercedes Grijalvo
Polytechnic University of Madrid, School of Industrial Engineering, Department of Industrial Engineering, Business Administration and Statistics, Spain

Miguel Blanco Callejo
Universidad Rey Juan Carlos, Spain

Esin Sadikoglu and Talha Temur
Gebze Institute of Technology, Turkish Standard Institution, Turkey

Alper Ertürk
Vlerick Leuven Gent Management School, Belgium

Printed in the USA
CPSIA information can be obtained
at www.ICGtesting.com
JSHW011442221024
72173JS00004B/912

9 781632 404183